Preface

This book has grown out of my notes compiled during the last 16 years while teaching economic entomology and conducting research in this field at the University of Ghana, Legon. The necessity for such a book has become especially apparent during the last five years when examining graduate and undergraduate students in African Universities and Colleges. African countries are currently emphasizing two aspects of agriculture: *plant production* and *plant protection*. Universities and Training Colleges are laying stress on principles and practice of modern agriculture in their curricula. The introduction of agriculture at 'A' level has added a new dimension to the rapid development of agriculture on the continent. However, there is no book available which might consider aspects such as pest control in the context of agriculture in Africa.

The study of pest control has grown vastly since the introduction of the first synthetic insecticide (DDT) in 1940. Indeed, most advances in our knowledge in this field have taken place in the last thirty years. The literature on the subject is expanding rapidly and a number of comprehensive books are currently on the market. But, because of their prohibitive prices, most of these are out of the reach of intended readers in developing countries and they lack relevance to agriculture in Africa. Further, a survey of 20 university and related libraries in East and West Africa revealed that 17 libraries had little or none of the pre-1975 literature cited in this book. And the three remaining libraries had barely half of the post-1975 literature.

This book is intended primarily to acquaint students of Agricultural Science and related subjects embarking upon diploma, undergraduate and M.Sc. courses in Africa with the fundamentals of pest control. Anyone with a fair general education, however, should profit from this book and it is hoped that it will be used by extension officers, 'A' level teachers, farmers and the general public interested in strategies of modern pest control. Whether in Africa or elsewhere, a knowledge of fundamentals is vital for the development of sound plant protection and pest control systems.

This is a textbook and, as such, not an exhaustive treatise. Books which review the subject of each section and subsection of the present volume are available. But an effort has been made to treat the subject matter in relation to the African continent. The material presented here is a bare summary of existing knowledge with a minimum of discussion. Entry into the literature, little known or not easily available to students in the developing world, is provided by the references. Some idea of the explosion of economic literature may be gained from a review of literature on the status of viruses pathogenic to insects and mites, where David (1975) (see p.145) had to consult over 400 papers published since January 1970. Thus the emphasis in this book on review articles at the expense of original works should be understood. The sheer volume of existing literature left little other alternative. In view of the rapid advances in economic entomology, a study of the latest issues of *Reviews of Applied Entomology*, *Annual Review of Entomology*, *Journal of Economic Entomology*, *Bulletin of Entomological Research*, *Tropical Pest Management* (formerly *Pest Articles News and Summaries, PANS*), *Insect Science and its Applications*, and other similar periodicals will still be necessary for the student who wants to keep himself up to date.

Acknowledgements

It is a pleasure to record my gratitude to a number of colleagues who have helped me, in various ways, in this venture: Mr. B. Hughes (Zoology Department, Legon) has read the book and made valuable suggestions and comments throughout and without his assistance this book would not be in its present shape; Professor M. Way (Imperial College, London) advised on the drafts of the book and encouraged me to send individual chapters to various specialists. I am most grateful to him, Mr. Hughes and to the following scientists who read and criticized various chapters in draft: Professor L. H. Rolston (Louisiana State University, Baton Rouge) – Chapters 1–7; Dr J. Wagge (Imperial College, London) and Dr R. I. Sailer (University of Florida, Gainesville) – Chapter 8; Dr G. H. S. Hooper

Insect Pest Control

with special reference to African agriculture

R. Kumar Ph.D (Raj.), Ph.D (Qld), F.R.E.S. (Lond.), F.E.S.I.

Professor of Entomology and Head, Department of Biological Sciences, Rivers State University of Science and Technology, Port Harcourt.

Formerly Professor of Zoology and Head, Department of Zoology, University of Ghana, Legon.

First published 1984 by
Edward Arnold (Publishers) Ltd
41 Bedford Square
London WC1 3DQ

British Library Cataloguing in Publication Data

Kumar, R.
Insect pest control.
1 Insect control
I Title
632'.7 SB931

ISBN 0-7131–8083–8

Printed and bound in Great Britain at
The Camelot Press Ltd, Southampton

(University of Queensland, Brisbane) – Chapter 9; Dr M. Elliott (Rothamsted Experimental Station, Harpenden) and Dr R. Loosli (CIBA-GEIGY Ltd, Basel) – Chapter 10; Dr G. A. Matthews (Imperial College, London), Drs R. Loosli, H. Zemp and O. Mueller (CIBA-GEIGY Ltd, Basel) – Chapter 11; Dr E. Bernays and Dr A. R. McCaffery (COPR, London) – Chapter 12; Dr R. I. Sailer – Chapter 13; these colleagues have frequently assisted me by supplying valuable reprints and information; they are of course in no way responsible for the views I have expressed or any errors and omissions which still remain. For the supply of valuable information and permission to refer to their work, I am also very grateful to: Dr E. Reye (University of Queensland, Brisbane) for the supply of Fig. 1 and permission to quote his work; Dr H. Wharton (C.S.I.R.O., Canberra) for permission to cite his work; Dr R. Passlow (Department of Primary Industries, Brisbane) for permission to refer to his work; Dr P. T. Walker (COPR, London) for the supply of information on pest forecasting services in Europe, U.S.A. and Japan; Dr L. A. Falcon (University of California, Berkeley) for permission to use Table 11; and to Professor T. R. Odhiambo for reading an early draft of the book and the supply of some invaluable literature.

For permission to reproduce illustrations and tables, thanks are due to the authorities of: International Institute of Tropical Agriculture, Ibadan (Tables 1 and 3); Commonwealth Agricultural Bureaux, Slough (Table 4); United States Council on Environmental Quality, Washington (Table 5); Blackwell Scientific Publications, Oxford (Table 5); Entomological Society of America, Maryland (Tables 7 and 9); Centre for Overseas Pest Research, London (Table 10; Figs 3 and 4, and safety precautions for pesticide application – p. 223–226); Rijksuniversiteit-Gent and Faculteit van de Landbouwwetenschappen, Gent (Fig. 2); C.S.I.R.O. (Australia) (Entomology Division) and Dr D. F. Waterhouse (Fig. 8a); British Crop Protection Council, Croydon (Fig. 9-chlordane; Fig. 10); Edward Arnold (Publishers) Ltd, London (Fig. 17); World Crops, Surrey (Fig. 20a), and Ciba Geigy Ltd, Basel (Figs 19, 20 b-c, 21). Grateful acknowledgement is also made to authors and publishers of original material which has been used to provide formulae etc. used in this book. Where possible full citations are given in the text.

The work was partly supported by grants from the Curriculum Division of Ghana Teaching Service and PAMOS (Ghana) Limited (BAYER, Agrochemicals Division). My wife, Dorcie, typed the first draft of the manuscript. Miss K. Woode prepared the index and Messrs. J. Allotey and J. Hama assisted in various ways. Criticism from the past students as well as generous facilities provided by the University of Ghana considerably assisted the completion of the book.

Legon, 1982 R. Kumar

Contents

GENERAL PRINCIPLES

1 Insects and Man in Africa

Insects are man's chief competitors on earth and to some extent his benefactors. They eat his crops and some of his other possessions. They transmit diseases to him and his domesticated animals. Insects are regarded as the most successful group within the Animal Kingdom; over 80% of all living animals are insects. About one million species of insects are known and over 6000 new species are described every year. As Professor Wigglesworth (1948) pointed out, insects are today, and probably through past ages always have been, the terrestrial animals most adaptable to changing food and climatic conditions and to competition with other animals. Except for the open oceans, insects are found in every conceivable habitat, from arctic snows to springs as hot as more than 40°C. A few species live in seemingly impossible environments, e.g. pools of crude petroleum and argol (containing 80% potassium bitartrate) (Imms, 1964). Prominent among the reasons for their success are: ability to live in and adapt to diverse habitats, high reproductive capacity, ability to consume different kinds and quantities of food, and ability to escape quickly from their enemies.

Insects are held responsible, at least partly, for the decline of the Golden Age in Greece and the fall of the Roman Empire. The early fall of the powerful and rich Mali Empire has been attributed to human trypanosomiasis – a disease that almost completely wiped out livestock production in the Mali Republic during the recent Sahelian drought (Odhiambo, 1977). Throughout history, as Cloudsley-Thompson (1976) says, 'insects have destroyed armies and confounded generals'. The dramatic failure of several expeditions to West Africa by conquering European adventurers has been attributed by historians to insect-transmitted diseases such as malaria, sleeping sickness, yellow fever, filariasis, river blindness, etc. The designation of the West Coast of Africa as the 'whiteman's grave' is no small tribute to these six-legged arthropods. Similarly, sixteenth century Portugese expeditions into the interior of East Africa floundered time and again when their horses and camels fell prey to sleeping sickness transmitted by tsetse flies. This same fly, which has aptly been called the 'bane of Africa', also prevented permanent settlement in the interior of tropical Africa by Arabs, the continent's first colonists. An estimated 10 million square kilometres of the African continent, most of it the better watered country and the more productive part, lying between 14°N and 29°S, is infested by tsetse flies (*Glossina* spp.) (Buxton, 1955). The activity of this fly in Nigeria necessitates the annual treatment of 400 000 – 750 000 head of cattle to control trypanosomiasis which these flies transmit. Odhiambo (1977) estimates that currently some 25–30% potential livestock production areas in Africa are not utilized because of livestock trypanosomiasis. Walsh (1968) estimates that as a result of river blindness, a disease carried by blackflies (*Simulium* spp.) 20 000 people in Nigeria are blind. At the moment we are waging an undeclared war against insects in a competitive struggle for existence. If man, by his folly of war and pollution, unwittingly exterminated himself there is little doubt that insects would take his place as the dominant terrestrial animals.

In the tropics, where climatic conditions are more conducive to their rapid multiplication, insects are all too familiar; from their bites and stings, the havoc they cause to crops, poultry and cattle, and the diseases they harbour and transmit. Hill (1975) has listed 407 insect species of major and 778 species of minor importance occurring on 48 major tropical crops. These figures do not include numerous other species recorded on these crops or the pests of stored products. For example, Lamb (1974), in listing the pests of tropical crops, states that no less than 4098 species of insects have been recorded on rice alone. Le Pelley (1973), in his review of coffee insects, says that more than 850 species of insects are known to feed on coffee, about 400 species in the Ethiopian region alone. Hargreaves (1948) listed 1326 species of insects on cotton and 482 of these are recorded from Africa south of Sahara.

Further, pests of crops and their status are likely to vary according to the local environmental conditions (Kumar, 1983; see next chapter for examples). Often estimates of crop losses due to pests are few and very scattered in the literature (Walker,

1975). In West Africa we are aware of the ravages caused by cocoa mirids to the cocoa tree, *Theobroma cacao*. Losses due to their activities in Ghana alone have been estimated at between 60 000 and 80 000 t compared to an actual annual crop of 200 – 250 000 t (Hale, 1953). Mirids are cocoa pests from Sierra Leone to Zaire, Uganda and the Malagasy Republic. Every year several West African governments spend large sums of money in trying to check the swollen shoot disease which is transmitted by mealybugs. Despite extensive research, the only method used to control this disease is the destruction of affected trees. Since 1946 well over 150 million diseased trees have been destroyed in the eastern region of Ghana alone, yet the method is only partially successful, adequate control is not being achieved and the incidence of the disease remains high (Legg and Kenten, 1968). This poses a very complicated financial and human problem. The farmer is told to destroy his trees that may be carrying cocoa pods and then he is asked to plant new cocoa trees again in the same place. Nigeria, Ivory Coast, Cameroun, Sao Thomé, Uganda and several other African countries suffer considerable losses of cocoa due to the activity of the insect pests of cocoa. Cocoa in West Africa is attacked by as many as 320 species of insects but most of these are kept in check by natural enemies. The use of pesticides in the attempted control of cocoa mirids has resulted in the killing of a number of non-target organisms which are parasites and predators on the pests. This has probably contributed to the emergence of a new pest of cocoa, the shield bug *Bathycoelia thalassina*, which is reported to cause an 18% loss in the yield of cocoa in the Eastern and Brong-Ahafo Regions of Ghana (Owusu–Manu, 1971). In Nigeria, Ivory Coast, Cameroun, Uganda and other African countries considerable sums are spent every year in an attempt to control various cocoa pests but no lasting control measures are yet available.

Closely related to cocoa is the kola tree (*Cola nitida* and *C. acuminata*) which is the main source of income to many farmers in the forest zone of West Africa. Daramola (1974), in a review of the pests of this crop, indicates that over 19 species of insect pests are of major importance though reliable data on the losses due to them are still lacking.

The coffee plant (*Coffea* spp.), a native of Africa now flourishes in many tropical countries especially in Central and

South America. More than 30 African countries already grow the crop. But wherever coffee is grown, it is liable to attack by additional insect pests, many indigenous to Africa. The coffee berry borer, *Hypothenemus hampei*, a native of Africa, has now been exported to nearly all parts of the tropics. In Brazil this pest has caused major crop losses and remains a most serious pest of coffee in that country (Le Pelley, 1973). All over Africa and elsewhere large sums of money are spent in controlling this and other pests (see Le Pelley, 1973 for a review of coffee insects).

The oil-palm industry in Western Ghana, Ivory Coast, parts of Nigeria, Sierra Leone, Dahomey and the Cameroun is seriously affected by the activities of a leaf-miner whose infestations may be so heavy that only the 'spikes' of the trees remain green. Attack by beetles may at times result in 70–90% loss to oil palms in Nigeria (Walker, 1975). Even the coconut palm, popularly thought to be relatively free from pests, may be attacked by no less than 23 species of insects in Ghana alone (Boakye, 1970).

Vast areas of farm land, forest and natural pasture in many African countries have been devastated by locust swarms. The African migratory locust, *Locusta migratoria migratorioides*, has been responsible for serious destruction of crops in Nigeria. According to Onazi (1968), £300 000 worth crops (mostly cereals) were destroyed by these insects between 1925 and 1934. In 1959 the cost of control operations against the migratory locust amounted to £14 000. MacCuaig (1963) stated that in Morocco, a major invasion by the locust, *Schistocerca gregaria*, caused damage estimated at over £4 000 000 within a few months in 1954–55. In 1958 Ethiopia lost 167 000 t of grain, enough to feed more than a million of her people for a year. In 1958 one swarm in Somalia occupied over 1000 km^2 and contained an estimated 40 billion locusts – a menace requiring 80 000 t of food a day (Conley, 1969). Walker (1967) estimates that Tanzania and Kenya lost £18 700 and £56 200 worth of cereals respectively to locusts in 1963. Dirsh (1965) includes over 500 genera and about 500 species from Africa in the superfamily Acriodoidea, many of which are without doubt potentially serious pests.

The grain legumes, such as cowpea (*Vigna unguiculata*), soybean (*Glycine max*) and pigeon pea (*Cajanus cajan*) provide green leaves, green pods and dry seeds which are important sources of palatable high quality protein in many African countries. One of the greatest limiting factors in attempts to increase the productivity of grain legumes is the wide range of insect pests with which they are associated. Over 130 pest species have been recorded on grain legumes in Africa and they may attack virtually every part of the crop – roots, stem, leaves, flowers and pods. Raheja (1976) estimated that in Northern Nigeria potential loss in yield of cowpea due to insects is over 90%, with 70% of this loss occurring during flowering and pod formation. Actually, if cowpea is grown as a single crop, it is virtually impossible to obtain any meaningful harvest without the use of insecticides to combat pests. Okigbo (1978) concludes that 'the potential of grain legumes will never be realized without understanding the pests and the pest control problem of legumes in the field, in transit, storage, processing and processed products and without developing scientific pest management systems for grain legumes'.

Cereal crops in Africa are liable to attack by a variety of insect pests. Walker (1967) in a survey of the estimated losses in yield in Tanzania and Kenya in 1963 attributed losses as follows:

- to stem borers, 18% in Kenya, 27% in Tanzania;
- to sorghum shootfly, 20% in Kenya, 4% in Tanzania;
- to *Heliothis*, 15% in Kenya and 3% in Tanzania; to barley fly, 14% in Kenya; to grasshoppers, 10% in Kenya, 7% in Tanzania;
- to cutworms, 2% in Kenya and 4% in Tanzania;
- to *Nematocerus*, 4% in Kenya; to white grubs, 2% in Kenya and 3% in Tanzania;
- to sorghum midge, 5% in Kenya;
- to sorghum sucking bugs, 6% in Tanzania; to black wheat beetle, 5% in Kenya;
- to *Gonocephalum*, 2% in Kenya;
- to *Epilachna*, 2% in Tanzania;
- to aphids, 2% in Kenya.

In other parts of Africa insect pests cause comparable losses to cereals (see Whitney, 1977 for details).

Root and tuber crops which are very important carbohydrate foods in the African tropics are attacked by at least 60 species of insect pests although few loss assessment studies have been made (see Whitney, 1977 for details). Yam tubers in Nigeria are liable to attack by 48 different species of insects (Toye, 1976). Yields of cassava may be reduced by as much as 80% by the whitefly-transmitted cassava mosaic disease (Beck, 1971). Damage to tubers by yam beetle ranges from 5% to 70% and is said to be always more than 20% in the main production areas (Whitney, 1977).

It was stated earlier that over 1000 species of insects have been recorded on cotton. Only a few of these are of economic importance but have resulted in low yields of seed cotton wherever the crop is grown without pest control (Tunstall & Matthews, 1972). Although a rather recent crop to be grown in several West African countries, this crop is seriously attacked by at least 9 species of insect pests. Among these bollworms may reduce cotton yields by 70–80%. Ripper & George (1965), in their book on *Cotton Pests of the Sudan*, state that damage by pests to cotton, unless controlled, 'would in many years reduce the cotton yield below the economic break-even mark and thus make cotton production unprofitable'.

The examples of crop losses due to insect pests given above may be multiplied: no field crop in Africa is free from attack, to some degree, from insect pests. For the entire world, the total value of losses in 1967, as a percentage of the potential crop, has been estimated at 14% due to insects, 12% to disease and 9% to weeds (Cramer, 1967). Summarizing over regions, Cramer estimates that Africa loses 9% of the total and 13% of the potential value of crops to insect pests. Recent studies indicate that these figures are likely to be gross under-estimates. It is unfortunate that such losses should occur in areas of the world where population growth and the resulting demand for food is the greatest.

The story does not end there. Whatever produce we are able to retrieve from the field must be stored and here insects come into the picture again. A large number of insects, including many species of beetles and moths, attack crops in farmers' bins, in mills, warehouses, retail stores and in the home. The damage done in this way is estimated to run into millions of dollars annually. Cornes (1963) lists 219 species belonging to 56 genera of 11 orders of insects associated with stored products in Nigeria. Hall (1970) stated that in some tropical countries, losses during storage, processing and marketing may be as high as 50%. Estimates of these losses from the time of harvest until consumption by man, vary from country to country and are available for only a few countries in Africa. In Ghana, out of a total annual harvest of 250–300 000 t of maize about 20% is lost to insect and rodent damage in storage (Prempeh, 1971). This means an annual loss equivalent to about 4–5 million cedis, (£2–2.5 million at 1971 values) (Prempeh, 1971). According to Ogunlana (1976), in Nigeria up to 27.7% losses have resulted from insect damage to maize stored in cribs for 4 months without pest control. Even at 4% level such wastage would amount to an annual loss of ₦5 000 000, currently about £4 000 000. Wheatley (1973) pointed out that for maize, direct and indirect farm losses in tropical countries vary from 23–35% leading to an overall loss of about 2 million tonnes annually in developing Africa. Caswell (1971) has estimated that 4.5% of annual production of cowpea is lost to insect pests in stores in Northern Nigeria. Cocoa in storage awaiting export is heavily attacked by insect pests and has to be protected by daily spraying with insecticides. Annual losses of dried fish by beetle pests amount to 12 million Naira (about £10 million) in Nigeria (Toye, 1976) while in Mali losses are worth about £1 500 000 (Aref *et al*., 1964). Dichter (1976) estimates that in sub-Saharan regions of Africa, losses of food grains during storage at farm or village level can amount to 25–40% of the harvested crop. This wastage of food cannot be tolerated when so many people in Africa remain hungry.

To the above losses should be added the economic damage caused by insects which act as vectors of debilitating diseases of man such as malaria, onchocerciasis, trypanosomiasis and other diseases of his animals. Mosquito-borne malaria is usually said to be the world's most devastating disease and is still credited as the biggest killer of mankind. Southwood (1977) points out that about one in six of mankind is suffering from insect-borne diseases. The cost in terms of prevention, control, treatment, human suffering and loss in productivity from these diseases are enormous. The long-term price of using insecticides becomes inestimable when the possible effects of pollution are considered. Pollution poses a problem to human health since some pesticides persist environmentally and are accumulated in our tissues with multiple, unpleasant and even lethal consequences. For this reason, some pesticides, such as DDT, have been banned from many countries. When an insecticide is first used it may be very effective but, over the years, becomes less so if insects become resistant to it. With the present rate of use, many insect pests may before long become resistant to all known pesticides. Some of the consequences of the use of pesticides to control insect pests have been outlined elsewhere (see Chapter 10 on chemical control). In the world, there are about 10 000 species of major insect pests and, as Odhiambo (1975) has noted, tropical Africa has the distinction of harbouring many of the major pests – the desert, migratory, red, and brown locusts, the African and spotted armyworms, numerous termites and ticks, and nearly 4000 other important insect pests of man, his crops and animals.

On the other hand, the beneficial value of insects in pollinating many crops is almost *inestimable*. Any agricultural practice which harms these insects is likely to cause a lowering in the yield of crops requiring pollination by insects. Silk produced by the silkworms (*Antheraea*, *Bombyx*, *Philosamia*) has been considered a major factor in encouraging travel and trade between East and West. The produce of honey bees (*Apis mellifera*) was a relatively important food prior to the availability of sugar from cane (Southwood, 1977). In some countries, such as Tanzania and Botswana, beeswax is still a major export crop. There are, too, numerous insect enemies of pests, and their propagation is essential to a balanced pest management programme (see Chapter 8 on biological control). One of the very important contributions of insects is their feeding on dead animal and plant matter, e.g. termites feed on dead wood, whilst ants and fly-larvae tunnel into animal carcasses and hasten their decomposition. Dung beetles feed exclusively on the dung of cattle and other ungulates and have become famous from the observations of Fabre. The value of insects in re-cycling dead animals and plants, returning organic material to the soil, is well known, and but for their activity, the world would perhaps be littered with refuse.

Being abundant, insects have proved to be important as experimental animals. Many discoveries in the biological sciences, particularly in genetics, were made with insects. Almost all significant basic concepts in transmission genetics were either developed from or verified by work on fruit-flies (*Drosophila* spp.). *Drosophila* was used as the basis for amplifying Mendelian genetics and giving it its present form (Brown, 1973). Our ideas on metabolic pathways, especially the cellular basis of oxidative respiration, fundamental to the supply of energy in living cells, were extended by work done on blowflies. Insects are currently engaging the attention of scientists from a variety of disciplines and have led to the present controversy over 'sociobiology' (Wilson, 1975). Insects, because of their many useful characteristics, have to-day become a valuable tool for the study of animal behaviour (Richard, 1973).

From this brief account it is clear that insects affect the welfare of human beings and their study involves potential economic and social benefits to man. No nation can afford to be complacent about the problems posed by insects. Here the role of entomologists becomes extremely important since it is they who are expected to manipulate and control insects in the biological environment. To do this, they need much more information than is at present available on the biology, ecology and behaviour of our pests. When this is available they may be able to formulate pests management systems. Nowhere is the need greater than in Africa which has too few entomologists for the many complex pest problems. For example, in Kenya only two entomologists are currently studying the pests of coffee which is grown over 200 000 ha of land. In Ghana, five entomologists are working on pest problems of cocoa which is grown over an area of 1.69 m ha. In Nigeria, four entomologists are employed on cocoa planted over 480 000 ha of land. Part of the problem is to educate administrators that, although costly, scientific research is a good investment and most productive when many scientists work simultaneously on the same problem. Toye (1976) estimates that presently there are no more than ten entomologists per 4.5 million people in Nigeria. The ratio elsewhere in Africa is even more lamentably inadequate. In 1979, on a continental population basis, Africa was estimated to have one entomologist to 608 000 head of population (Odhiambo, 1981).

1.1 Literature cited

Aref, M., Timbley, A., & Daget, J. (1964). Fish and fish processing in the Republic of Mali. 3. On the destruction of dried fish by Dermastid insects. *Alex. J. Agric. Res.*, **12**, 95–108.

Beck, B. D. A. (1971). Cassava production in West Africa. *Ford Foundation IRAT/IITA seminar on root and tuber crops in West Africa, Ibadan, 22nd–26th February, 1971.* 11pp.

Boakye, D. B. (1970). In *Coconut in Ghana. Bulletin C.R.I., C.S.I.R.*, **3**, 13.

Brown, S.W. (1973). Genetics – the long story. pp. 407–32. In *History of Entomology.* Eds. Smith R. F., Mittler, T. E. and Smith, C. N. *Annual Reviews Inc.*, 4139 EL Camino Way, Palo Alto, California and *Entomological Society of America, Maryland*. 517 pp.

Buxton, P. A. (1955). *The Natural History of Tsetse Flies*. H. K. Lewis, London. 816 pp.

Caswell, G. H. (1971). The impact of infestation on commodities. *Ford Foundation IRAT/IITA seminar on grain storage in the humid tropics, Ibadan, 26–30th July, 1971.*

Cloudsley-Thompson, J.L. (1976). *Insects and History*. Weidenfeld and Nicolson, London. 242 pp.

Conley, R. A. M. (1969). Locusts: teeth of the wind. *National Geographic Magazine*, **136(2)**, 202–26.

Cornes, M. A. (1963). Further investigations into the small scale storage of maize in cribs. *Nigerian Stored Products Research Institute, Annual Report*, 101–10.

Cramer, H. H. (1967). Plant protection and world crop production. *Pflanzenschutz Nachrichten, Bayer*, **20**, 1–524.

Daramola, A. M. (1974). A review of the pests of *Cola* species in West Africa. *Nigerian J. Ent.*, **1(1)**, 21–9.

Dichter, D. (1976). The stealthy thief. *Ceres, FAO Rev.*, **9(4)**, 51–3, 55.

Dirsh, R. M. (1965). *The African Genera of Acridoidea*. Cambridge University Press and the Anti-Locust Research Centre, London. 579 pp.

Hale, S. L. (1953). World production consumption – 1951 to 1953. *Rep. Cocoa Conf. Lond.*, 1953, 3–13.

Hall, D. W. (1970). *Handling and Storage of Food Grains in Tropical and Subtropical Areas*. FAO, Rome. 350 pp.

Hargreaves, H. (1948). *List of Recorded Cotton Insects of the World*. Commonwealth Institute of Entomology, London. 50 pp.

Hill, D. (1975). *Agricultural Insect Pests of the Tropics and their Control.* Cambridge University Press, London. 516 pp.

Imms, A. D. (1964). *Outlines of Entomology*. 5th ed. Methuen, London. 224 pp.

Kumar, R. (1983). Agricultural pests of crucial economic importance in the tropics and their control. *Current Themes in Tropical Science, Pergamon Press Ltd* **3**, (in press.)

Lamb, K. P. (1974). *Economic Entomology in the Tropics*. Academic Press, London. 195 pp.

Legg, J. T. & Kenten, R. H. (1968). Some observations on cocoa trees tolerant to cocoa swollen shoot virus. *Tropical Agriculture*, **45(1)**, 61–5.

Le Pelley, R. H. (1973). Coffee insects. *Ann. Rev. Ent.*, **18**, 121–42.

MacCuaig, R. D. (1963). Recent developments in locust control. *World Rev. Pest Control*, **2(1)**, 7–17.

Odhiambo, T. R. (1975). This is a dudu world. *First Lecture*, 4th June 1975, ICIPE, Nairobi, 3–17 pp.

Odhiambo, T. R. (1977). Entomology and the problems of the tropical world. *Proc. XV. International congr. Ent. Washington*. 52–9.

Odhiambo, T. R. (1981). Insect pests. *Proceedings of the Symposium on the state of biology in Africa. International Biosciences Networks, Accra, Ghana*., April 1981, 112–23.

Ogunlana, M. C. (1976). Opening address at the 9th Annual Conference and 10th Anniversary celebrations of the entomological Society of Nigeria. *Proc. 9th Ann. Conf. and 10th Anniv. Celeb. ent. Soc. Nig*., 18–23.

Onazi, O. C. (1968). Locust control in Nigeria. *Fifty years of Applied Entomology in Nigeria . Proc. ent. Soc. Nigeria*, 62–7.

Okigbo, B. N. (1978). Grain legumes in the Agriculture of the tropics. p. 1–11. In *Pests of Grain Legumes: Ecology and Control*. Eds Singh, S. R., van Emden, H. F., and Taylor, T. A. Academic Press, London. 454 pp.

Owusu-Manu, E. (1971). *Bathycoelia thalassina* – another serious pest of cocoa in Ghana. *C.M.B. Newsletter, Accra*., **47**, 12–14.

Prempeh, H. R. B. A. (1971). Maize crop harvesting, processing and storage in Ghana. *'Maize the wonder crop', Symposium at UST, Kumasi*. 81–7.

Raheja, A. K. (1976). Assessment of losses caused by insect pests to cowpeas in Northern Nigeria. *PANS*, **22(2)**, 229–33.

Richard, G. (1973). The historical development of nineteenth and twentieth century studies on the behaviour of insects. p. 447–502. In *History of Entomology*. Eds Smith, R. F., Mittler, T. E., and Smith, C. N. *Annual Reviews Inc*., 4139 EL Camino Way, Palo Alto, California and *Entomological Society of America, Maryland*. 517 pp.

Ripper, W. E. & George, L. (1965). *Cotton Pests of the Sudan*. Blackwell Scientific Publications, Oxford. 345 pp.

Southwood, T. R. E. (1977). Entomology and mankind. *Proc. XV. International Congr. Ent. Washington*., 36–51.

Toye, S. A. (1976). The role of entomology in the economic development of Nigeria. *Proc. 9th Ann. Conf. and 10th Anniv. Celeb. ent. Soc. Nig*., 3–10.

Tunstall, J. P., & Matthews, G. A. (1972). Insect pests of cotton in the old world and their control. p. 46–59. In *Cotton*. CIBA-GEIGY

Agrochemicals. *Technical Monograph No.* **3**. 80 pp.
Walker, P. T. (1967). A survey of losses of cereals to pests in Kenya and Tanzania. Paper presented in *FAO Symposium on Crop Losses, Rome*, 79–88.
Walker, P. T. (1975). Pest control problems (pre-harvest) causing major losses in world food supplies. *FAO Plant Protection Bulletin*, **23**, 70–7.
Walsh, J. P. (1968). The Simuliidae and their control with special reference to the Kainji control scheme. *'Fifty years of applied entomology in Nigeria'. Proc. ent. Soc. Nigeria*, 1968, 100–6.
Wheatley, P. E. (1973). Post harvest deterioration – the maize storage problem in less developed countries of Africa. *Chem. & Industry*., 1049.
Whitney, W. K. (1977). Insect and mite pests and their control. pp. 195–235. In *Food Crops of the Lowland Tropics*. Eds Leakey, C. L. A., and Wills, J. B. Oxford University Press, Oxford. 345 pp.
Wigglesworth, V. B. (1948). The insect as a medium for the study of physiology. *Proc. R. Soc. Lond.* (B)., **135**, 430–46.
Wilson, H. O. (1975). *Sociobiology*. The Belknap Press of Harvard University Press, Cambridge, Massachusetts. 697 pp.

2 When is an Insect a Pest?

2.1 Definition of a pest

The Concise Oxford Dictionary defines pest as, 'a troublesome or destructive person, animal or thing'. Pest comes from the Latin, *pestis* meaning plague and it is often loosely used. In pest control it is most important to determine whether, when and how an animal or plant is a pest or vector. Vector is a carrier of infection and is derived from the Latin, *vehere vectum* – meaning convey.

Williams (1947) states, 'an insect pest is any insect in the wrong place – from a human point of view; just as a rose bush is a weed when it is growing in a cabbage patch'. He has given an interesting example of how an insect may be a pest under one set of conditions but not under another set. Thus the painted lady butterfly (*Pyrameis cardui*) in Britain is merely a 'thing of beauty and joy for ever'; but in France it is a pest of artichokes, and in North America it acts as an agent of biological control of the troublesome thistle weeds. Thus from the human angle this butterfly is only a pest in France.

In West Africa, cassava plant (*Manihot esculenta*) is attacked by the variegated grasshopper (*Zonocerus variegatus*). Experiments in Nigeria have shown that defoliation of cassava after 9–11 months has no effect on the yield of the plant but at 7 months, and especially if defoliation was repeated, the mean yield is reduced to about 60% (Anonymous, 1974). Should the grasshopper be called a pest only when it attacks cassava plants less than 7 months old, and then repeatedly? A Fijian midge (*Culicoides belkini*) breeds only in brackish water and was not a known pest because few people built villages within reach of the midge. Many people, especially women, are allergic to the bites of this midge. But when swamp was reclaimed for urban development, bringing *Homo sapiens* within reach of the midge, the latter was immediately classified as a pest. Clearly, expansion of human habitation caused the emergence of a new pest.

Mosquitoes (*Anopheles* spp.) are well known vectors of malaria. Currently research is underway to develop a vaccine against malarial parasites. Should this be successful, would mosquitoes still qualify as pests?

Prickly pears (*Opuntia* spp.) are decorative garden plants in many parts of the world. But its members may multiply to such an extent that control measures have to be used against them. Prickly pears (of American origin) had occupied some 4 million hectares of land in Queensland, Australia by 1900 and 24 million hectares by 1925. Half of the area was so densely infested that it was useless for agriculture. The weeds were eventually brought under control by the introduction of a moth, *Cactoblastis cactorum*, from America. It is clear that the prickly pears may be a pest in one situation but not so under another.

Various definitions of a pest have been proposed and generally they have tended to be biased by the profession of the author:

> 'We can define pests as those injurious or nuisance species, the control of which is felt to be necessary either for economic or social reasons' (Clark, 1970).

> 'By pest, we mean any plant or animal species that is unacceptably abundant' (Bisplinghoff & Brooks, 1972).

> 'In all farming systems, many sorts of organisms compete with man at both the primary and secondary stages of production. They include insects, mites, and other arachnids, ticks and other exoparasites of animals; nematodes and other harmful parasitic worms of both plants and animals; fungi; bacteria; viruses; higher plants (weeds and poisonous plants); birds; and mammals. From man's point of view these, and the vectors that transmit some of them are harmful organisms, and in this Symposium we refer to them collectively as pests' (Bunting, 1972).

> 'Pest refers to all types of biological factors that reduce crop income: insects, weeds, disease, nematodes etc' (Carlson, 1973).

> 'Any animal which does economic damage to crops and domesticated animals, or is harmful to human health, constitutes a pest. Generally, pests have to be abundant, at least

at certain times, in order to do economic damage. There are exceptions to this however, and some species, particularly those carrying disease, can do considerable harm at very low densities' (Dempster, 1975).

Some countries have legislation defining the term 'pest'. For example, under the United States Federal Insecticide, Fungicide, and Rodenticide Act, the term 'pest' means (1) any insect, rodent, nematode, fungi, weed, or (2) any other forms of terrestrial or aquatic plant or animal life or virus, bacteria or other micro-organisms (except viruses, bacteria or other micro-organisms on or in living man or other living animals) which the administrator declares to be a pest . . . (section 2(t)). In Africa, such legislation, is as yet lacking.

It is clear from the foregoing discussion that pests are those species which by their activities become inimical to the welfare of man. Pests include some vertebrates such as certain birds and rodents; many species of insects, ticks, mites and other arachnids; nematodes and other parastic worms; weeds and other undesired plants; fungi, bacteria, viruses and other harmful micro-organisms. The present book deals only with the control of insects pests in agriculture.

From an agriculturalist's point of view the definition of an insect pest or vector is essentially economic, that is, whether or not an insect is causing sufficient damage to necessitate control measures. This idea of economic threshold was emphasized by Stern *et al*. (1959) who defined it as 'the density at which control measures could be determined to prevent an increasing pest population from reaching the economic injury level'. Recently, a concise definition by Carlson (1973) terms economic thresholds as 'levels of pest damage which call for the use of plant protection measures'. Thus the pest or vector in agriculture is an economic pest or vector.

The term 'sufficient damage' mentioned above is admittedly vague and perhaps too restrictive. Houseflies, for example, are considered to be pests by virtually everyone concerned with horses and cattle. Yet attempts to control them are rare because there is no satisfactory control. Further, within agriculture, insects can and do cause monetary loss below the economic threshold. Treatment may cost more than the return, but nevertheless damage is sustained and the cause of the damage is certainly a pest. How much damage or harm by a pest or vector is acceptable would have to be worked out for each particular situation. For example, 10% crop losses may be acceptable to a particular country or group of people who do not wish to use insecticides in controlling the damage in order to prevent or minimize injury to their environment. On the other hand a country with food shortages may opt to adopt control measures in order to harvest as much of the crop as possible and may even decide to control pests causing damage below economic thresholds. Economic thresholds will be dealt with more fully in the next chapter.

2.2 Build up of pest populations

Insects become pests as the result of a wide variety of factors and often an understanding of these will lead to improved systems of management. However, we must first briefly consider the ecosystems of which insects in nature form part, and the man-made agro-ecosystems derived from these in which insects in ecological disequilibrium become pests.

2.2.1 Ecosystem

An ecosystem comprises complexes of plants and animals together with the environmental factors which affect them. Each species lives in a particular environment comprising biotic and abiotic factors which have brought about its adaptation by evolution. The organisms tend to live in a dynamic equilibrium which is a self-regulating and self-maintaining system. Tansley (1920) in his concept of ecosystem, envisaged communities of all ranks to be the result of two processes, one set tending to maintain a *status quo*, the other set tending to destroy it. A range of diverse natural ecosystems exist. Some are more or less permanent while others may be temporary. For our purposes the following two categories of ecosystems will be considered:

Forest ecosystem Forest is a complex vegetation which usually forms a canopy beneath which are shorter trees, shrubs, climbers and herbs but from which grass is absent. The canopy may be so dense as to cut off as much as 99% of the incident sunlight, leaving little to reach the forest floor.

Savanna ecosystem Savanna is a type of vegetation consisting predominantly of grasses, burnt annually in the tropics, with other herbs, trees or woody plants which, however, do not form a continuous canopy. A large percentage of the incident sunlight reaches the ground and determines the fauna and flora which are so different from those of forest. In savanna, it rains 3–4 months a year while in forests it rains appreciably, usually in almost all months. Seasons are more marked in savanna than in forest.

Because of the complex stratification of the forest, there are more niches or 'job opportunities' available and therefore a greater number of species of animals are found in forest compared to the structurally simpler savanna ecosystem. The shade of the forest creates a micro-climate in which a large number of micro-habitats are formed. The great diversity of animal life there is related to the great diversity of plant life. Savanna contains relatively fewer species of animals which are bigger and lighter in colour than forest species. The forest ecosystem is stable seasonally and perennially. In conditions of environmental stability, more species survive. In such conditions species become more specialized and niches fragmented in accommodating the increased number of species. This small niche dimension is perhaps the reason that it is difficult to introduce a new animal in a forest ecosystem.

The savanna ecosystem is unstable seasonally, less diverse in numbers of species, but each species is represented by large populations which come into constant interaction with each other. Pest situations are therefore more likely to arise in savanna than in the forest.

Perturbation of ecosystems Consequences of introductions of animals into a new environment are depicted in Fig. 1. It is well to keep in mind, with regard to this diagram, than an animal or plant which takes over an unoccupied niche may have new ecological effects depending on the ecosystem.

2.2.2 Agro-ecosystem and development of pests on a crop

Agro-ecosystems are always of limited duration, mostly lasting a few months after planting, as in the case of many non-tree crops. It is well said that when man made the agro-ecosystems he also made pests. Modern agriculturists farm large areas of a

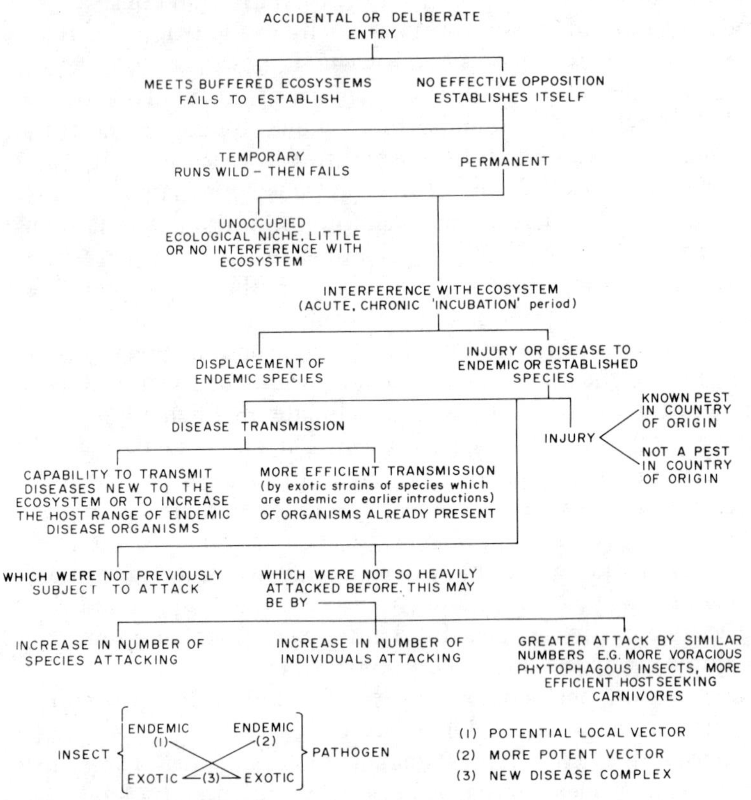

Figure 1 Consequences of the introduction of an organism into a new environment (after Reye, 1966, modified).

single crop plant. This facilitates planting, cultivating and harvesting. Any other neighbouring plant is unwanted and is usually destroyed. Thus the natural balance of animal and plant populations, the result of prolonged periods of evolution, is upset. The usual result is a drastic disturbance of the environment and a radical reconstruction of the faunas injurious to useful plants. Changes brought by modern agricultural practices may be summarized as follows:

Monoculture Cultivation of single crops on very large areas has created conditions favourable for specialized insect species to flourish and become notorious pests. Uvarov (1967) has summarized the effects of monocultures on certain tropical and subtropical crops. He has shown how some of the well-known pests of crops such as sorghum, maize and sugar-cane exist in the wild as feeders on wild grasses and are too few to be noticed. However, whenever these crops are grown on a large scale, they are infested to pest proportions. Monocultures have led to the simplification of fauna and flora thereby providing suitable habitats for a few species which specialize on particular crops. Uvarov (1967) has drawn attention to the work of Russian entomologists who carried out quantitative studies of insect populations of virgin grass steppe and of new wheat fields in Kazakstan. One of their important findings was that there was a greater specific variety of the insect populations of virgin steppe and hence a better ecological balance between its species. On the wheat fields on the other hand, there were a few insect species which soon became the worst pests of wheat.

Way (1971) has argued that we must not assume that large losses from pests are caused by crops being grown as a monoculture. In his opinion the large areas of monocultures may be a 'favourable characteristic merely through a simple diluting effect on colonising pests'. He has cautioned against the assumption that diversity in agro-ecosystems is desirable for minimizing pest damage. Taylor (1972) has attributed the more serious insect pest situations in Southern Nigeria to a greater diversity of ecosystems with more numerous wild host plants. Way (1976) states that 'the recent upsurge of *Heliothis armigera* and *Cryptophlebia leucotreta* as major pests of cotton in Northern Nigeria is attributable directly to increased diversity from the growing of maize and tomatoes which bridge gaps in their host plant sequences'. Although it is generally recognized that in natural ecosystems greater diversity leads to a greater chance of stability (Southwood & Way, 1970; van Emden & Williams, 1974 etc.), Way (1966; 1976) concludes the right kind of diversity is fundamental to pest control. He believes that 'irrespective of complex food web interactions, a few links in the web, perhaps only one, may be crucially important in affecting the abundance of a particular pest'.

Quality and quantity of food supply Man selects his crops for certain features. Generally they are high yielding varieties, have larger seeds and fruits which are more nutritious than in their wild progenitors. Changes in quality brought about by cultivation are generally believed to favour the pest (Strickland, 1960; van Emden, 1966). Further, large numbers of plants of the same species supply virtually unlimited food to a phytophagous insect which is thus able to multiply rapidly without restriction and with the minimum of exposure to the hazards of dispersal (Southwood & Way, 1970). Many pests of stored products exist at insignificant levels in the field but storage of food on a large scale results in their emerging as pests.

Host/natural enemy relationship The effects of parasites, predators and disease in keeping their hosts under control are now well recognized. However, modern agricultural practices, e.g. the use of pesticides, tend to upset this balance. Field experiments in Ghana have shown (Owusu-Manu, 1976) that Gamma-BHC, the commonly used anti-mirid insecticide in cocoa-farms, has an adverse effect on the natural enemies of a shieldbug, *Bathycoelia thalassina*, which until recently was seldom noticed on cocoa. He found that only 2% of the initial population of this insect reached the adult stage when they were exposed to natural enemies in the wild, but 71% reached adulthood when most of these enemies had been removed by spraying. It was concluded that the extensive use of lindane was responsible for *B. thalassina* reaching pest level. Similarly in Kenya, the widespread use of residual insecticides in coffee estates for the control of leaf miners (*Leucoptera* spp.) and other pests has reduced natural enemy populations of the giant looper, *Ascotis selenaria reciprocaria* and caused the latter occasionally to become a serious pest rather than a minor one. Kayumbo (1977) cites investigations in Tanzania to the effect that increasing use of insecticides at the Ukiruguru cotton research station appears to have reduced parasitism of *Heliothis* from about 30% to 5% between 1963 and 1973, presumably through destruction by insecticides. Pest outbreaks following the use of pesticides have been discussed by Coaker (1977).

There is further evidence that some predators on pests, due to their slow rate of increase, are often unable to increase as rapidly as their fast breeding hosts which therefore reach pest proportions. Reviewing outbreaks of forest insects, Voûte (1946) considered that phytophagous insects in a natural habitat are usually well regulated by natural enemies, especially polyphagous predators. When the density of a phytophagous insect exceeds a threshold level, due to some changes in their environment, or other factors, the insect can 'escape' from predation pressure and increase to pest proportions. Thus the 'escape' is considered by Voûte (1946) to be responsible for many severe pest outbreaks spread over many generations. Recently Itô and Nagamine (1981) have suggested that escape from predation may be the main cause of the maintenance of extraordinary high densities of *Mogannia minuta* (Cicadidae) on sugar-cane in Okinawa.

Introduction to new environments With the quick and efficient modern means of international transport, introduction of animals and plants from distant lands has become relatively easy. We are aware of the past experiences of the introduction of pests from their original homes to new environments, e.g. the accidental introduction of gypsy moth from Europe to North America, introduction of cattle tick into Australia and the import of foot and mouth disease of cattle into Britain by infected meat. According to Wharton (1966), the majority of Australian pests are exotic and have been introduced over the past 190 years. Within an ecosystem there is a well established system of natural enemies but a new environment may be without much check to the rapid growth of a pest population. Similarly, when crops are introduced into new environments, local insects may find them more suitable food plants and become pests on them. For example, the cocoa tree, *Theobroma cacao*, introduced from South America to West Africa and a backbone of the economy of several African countries, is attacked by no less than 320 species of local insects. The depredations of some of these (e.g. mirids and mealybugs) decrease cocoa yield by over 30% annually in Ghana.

Uvarov (1967), in discussing the formation of pest fauna of various crops, states, 'it is certainly a general rule that the injurious fauna of a given crop in a given country is produced as a result of segregation of a reduced number of species already present in wild habitats, and pest status is reached by them owing to improved conditions of existence'. However, Pemberton & Williams (1969) while considering the origins of sugar-cane pests state that pest faunas of sugar-cane, and perhaps of all crops, are of dual origin, comprising indigenous as well as introduced species from other countries. The proportion of each would however, as noted by these authors, vary considerably according to particular local situations.

The transport of pests and disease from country to country along with plant material or otherwise is now fully appreciated. Most countries now have legislation regulating plant import and adoption of measures, such as plant inspection, plant quarantine etc. Indeed, the first legislation designed to prevent introduction of foreign insects was enacted in Germany in 1873 against the grape root aphid (David, 1949). Where well enforced, such measures are known to have either delayed the spread of pests or totally precluded their entry into a country. Pemberton & Williams (1969), in their account of the origins and spread of sugar-cane pests, state, 'traffic of sugar-cane varieties, and without the phytosanitary precautions which are now regarded as essential, was responsible for spread of many sugar-cane insects, some of which today rank as major pests. The further spread of cane pests cannot be precluded completely'.

In addition to the ecological considerations discussed above, economic realities of modern agriculture, which is an expensive undertaking, should also be borne in mind. Purchase and maintenance of machines such as tractors and harvesters is costly in itself. And vital inputs such as fertilizers and water for irrigation must be provided. In terms of overall costs and returns, small crop losses are tolerable when investments are low but they would become unbearable when expenditures are high. This means that there is a tendency to set lower economic thresholds for high input crops, and insect damage which would otherwise be ignored is likely to become important and attributable to the presence of pests in these crops.

Food production hardly keeps pace with the current rapid increase in human populations. Therefore there is an unrealistic emphasis on eradicating any animal or plant that might lessen the yield. We are aware that in China to obtain maximum harvests, bird populations feeding on cereal crops were targeted for total eradication. These efforts only led to the emergence of secondary insect pests.

2.3 Variation in status of pests of crops

World agriculture is characterized by a diversity of its crops which must be attributed to particular environmental conditions prevailing in different parts of the world. Thus factors such as solar radiation, temperature, day length and rainfall interact in different ways in the monsoon, dry, subhumid, humid and other regions. The elements mentioned constitute the climate of a place and along with soil conditions, natural vegetation and local fauna determine the crops that can be grown in that place. Further, in the tropics, inadequate resources and the generally poor education of the farmers greatly influence the practice of agriculture. Unlike their temperate counterparts they are usually unable to combat pests.

Thus it will be clear that pests of crops and their status are likely to vary according to the local environmental conditions. Thus, the pests of cocoa are quite different in West Africa, South America and South East Asia. Even within West Africa the pest complex varies from country to country (Kumar, 1983). Similarly, excepting the pantropical green scale (*Coccus viridis*), major pests of coffee are seldom the same where coffee is grown in the tropics. The same diversity of pests is found on tea, cotton, sugar-cane, rice, coconut and other palms. There are, no doubt, exceptions to this statement: for example, the sorghum midge, *Contarinia sorghicola*, attacks sorghum wherever the crop is grown. The green vegetable bug, *Nezara viridula*, is generally agreed to be a cosmopolitan pest on cotton, tomato and castor; *Heliothis armigera* is a cosmopolitan pest, in the old world, of tobacco, sorghum, cotton, groundnuts, okra and pulses. Kumar (1983) lists 43 cosmopolitan and pantropical pests of crops in the tropics. Where the same pest occurs widely its pest status is not always the same. Damage inflicted varies according to the country, crop and season as well as the many other factors, including farming practices.

Some insect pests have increased in importance in the recent years. For example, the brown plant-hopper, *Nilaparvata lugens*, a minor pest in the past, has become a major pest of rice throughout tropical Asia. Losses due to this pest have been estimated at about US $300 million in 11 countries (Dyck, 1977). The whorl maggot, *Hydrellia* sp., once a pest of little significance, has become a major pest of rice in Philippines. The variegated grasshopper, *Zonocerus variegatus*, once of little consequence to Nigerian agriculture, has in the recent years become a major crop pest. Clearly an insect's status as a pest may change in response to changing conditions. The same insect may not be a pest throughout its range and may not consistently be a pest in the same locality. The basic causes of changes in pest status of insects are really not well understood in many cases.

2.4 Summary

Animals and plants are classed as pests when they cause sufficient damage to warrant application of control measures. Pest of crops in a given environment may be produced as a result of a combination of ecological factors. Some of these include selection of a species already present in the wild or due to accidental or deliberate introductions into new environments; modern agricultural practices such as monoculture and the use of pesticides; change in host/natural enemy relationship etc. They attain pest status because of the unlimited availability of a preferred host. The pests of crops and their status vary according to local environmental conditions. Little is known, in many cases, as to why their importance differs locally.

There are over a million species of insects of which perhaps less than 10 000 merit a pest status of some kind. Pest control systems should therefore be directed towards reducing the numbers of pest species below the threshold of economic importance. They would still be with us but not above the level that would cause significant economic damage. This would mean that there would be little or no interference with the ecosystems such as would be required by the unrealistic aim of total eradication.

2.5 Literature cited

Anonymous (1974). Control of *Zonocerus variegatus* L. in Nigeria. Interim Report. Jan. 1972–July, 1973. *Centre for Overseas Pest Research, London,* 1–17.

Bisplinghoff, R. L. & Brooks, J. L. (1972). Role of basic research in implementing pest control strategies. pp. 36–43. In *Pest Control Strategies for the Future*. National Academy of Sciences, Washington D.C. 376 pp.

Bunting. A. H. (1972). Ecology of agriculture in the world of today and tomorrow. pp. 18–35. In *Pest Control Strategies for the Future.* National Academy of Sciences, Washington, D.C. 376 pp.

Carlson, G. A. (1973). Economic aspects of crop loss control at the farm level. In *Crop Assessment Methods*. F.A.O. Manual on the evaluation and prevention of losses by pests, diseases and weeds (1977 supplement, not sequentially paginated).

Clark, L. R. (1970). Analysis of pest situations through the life system approach. pp. 45–58. In *Concepts of Pest Management*. Eds Rabb, R. L., and Guthrie, E. F., North Carolina State University., Raleigh, U.S.A. 242 pp.

Coaker, T. H. (1977). Crop pest problems resulting from chemical control. pp. 313–28. In *Origins of Pest, Parasite, Disease and Weed Problems.* 18th Symposium of the British Ecological Society, Bangor, 12–14 April, 1976. Ed. Cherret, J. M., and Sagar, G. R. Blackwell Scientific Publications, Oxford. 413 pp.

David, W. A. L. (1949). Air transport and insects of agricultural importance. *Commonwealth Institute of Entomology, London.* 11 pp.

Dempster, J. P. (1975). *Animal Population Ecology*. Academic Press, London. 155 pp.

Dyck, V. A. (1977). Paper presented in the Brown Planthopper symposium. April 18–22, 1977. *Int. Rice Res. Inst., Los Baños, Philippines*. Mimeographed.

van Emden, H. F. (1966). Plant insect relationship and pest control. *Wld. Rev. Pest Control*, **5**, 115–123.

van Emden, H. F., & Williams, G. F. (1974). Insect stability and diversity in agroecosystems. *Ann. Rev. Ent.*, **19**, 455–75.

Itô, Y., & Nagamine, M. (1981). Why a cicada, *Mogannia minuta* Matsumura, became a pest of sugarcane: a hypothesis based on the theory of 'escape'. *Ecological Entomology*, **6(3)**, 273–83.

Kayumbo, H. Y. (1977). Ecological background to pest control in mixed cropping systems. In *Proceedings of the Workshop on Cropping systems in Africa, Morogoro, Tanzania*, 1–6, December 1975, 51–3.

Kumar, R. (1983). Agricultural pests of crucial importance in the tropics and their control. *Current Themes in Tropical Science. Pergamon Press Ltd.*, **3**, (in press).

Owusu-Manu, E. (1976). Natural enemies of *Bathycoelia thalassina* (Herrich-Schaeffer), a pest of cocoa in Ghana. *Biol. J. Linn. Soc*., **8(3)**, 217–44.

Pemberton, C. E. & Williams, J. R. (1969). Distribution, origins and spread of sugarcane insect pests. pp. 1–9. In *Pests of Sugarcane*. Eds Williams, J.R., Metcalfe, J. R., Mungomery, R. W., and Mathes, R. Elsevier Publishing Company, Amsterdam. 568 pp.

Reye, E. J. (1966). Exotic insects and insect-borne diseases. Historical aspect. *News Bull. Ent. Soc. Old.*, **28**, 2–4.

Southwood, T. R. E. & Way, M. J. (1970). Ecological background to pest management. pp. 6–29. In *Concepts of Pest Management*. Eds Rabb R. L. & Guthrie, F. E. North Carolina State University, Raleigh, U.S.A. 242 pp.

Stern, V. M., Smith, R. F., van den Bosch, R. & Hagen K. S. (1959). The integration of chemical and biological control of the spotted alfalfa aphid. Part I. The integrated control concept. *Hilgardia*, **29**, 81–101.

Strickland, A. W. (1960). Ecological problems in crop control. In Wood, R. K. S. (ed.): Biological problems arising from the control of pests and diseases. *Symp. Inst. Biol.*, **9**, 1–15.

Tansley, A. G. (1920). The classification of vegetation and the concept of development. *J. Ecol.*, **8**, 118–49.

Taylor, T. A. (1972). Integrated control of plant pests in Nigeria. *Proc. FAO Panel on Integrated Pest Control 1972.*

Uvarov. B. (1967). Problems of insect ecology in developing countries. *PANS*, **13**, 202–13.

Voûte, A. D. (1946). Regulation of the density of the insect-populations in virgin-forest and cultivated woods. *Archives Neerlandaises de Zoologie*, **7**, 435–70.

Way, M. J. (1966). The natural environment and integrated methods of pest control. *J. Appl. Ecol.*, **3** (Suppl.), 29–32.

Way, M. J. (1971). A prospect of pest control. *Inaugural Lecture, Imperial College of Science and Technology, University of London*, 127–62.

Way, M. J. (1976). Diversity and stability concepts in relation to tropical insect pest management. *Proc. 9th Ann. Conf. and 10th Anniv. Celeb. ent. Soc. Nig.*, 68–93.

Wharton, R. H. (1966). Exotic insects and insect-borne diseases. Introduction. *News Bull. Ent. Soc. Old.*, **28**, 2.

Williams, C. B. (1947). The field of research in preventive entomology. *Ann. appl. Biol.*, **34(2)**, 175–85.

3 When to Control a Pest

3.1 Economic threshold

In order to consider using economically feasible control measures against pests, reliable information is needed on yield losses as a result of pest attack. For this purpose a knowledge of economic thresholds is essential. Economic thresholds are defined as levels of pest damage which warrant the use of plant protection measures (Carlson, 1973, Chiang, 1979). Two terms frequently used while considering control measures are defined by Stern *et al.* (1959) as follows:

Economic-injury level: the lowest population density which will cause economic damage.

Economic damage: the amount of injury which will justify the cost of artificial control measures.

A number of other technical terms appear in literature dealing with assessment of economic crop losses in yield caused by pests, and for annual crops these have been reviewed by Judenko (1972). The ability to determine an economic threshold of an insect pest on a crop is dependent on distinguishing the different infestation levels and the degree to which each level influences the harvested crop (Stern, 1973). According to Talpaz & Frisbie (1975) a threshold is a dynamic measure which may vary with infestation level, value of yield, cost of control and time of assessment. In practice the problem simply boils down to the fact that we should be able to obtain an accurate estimation of pest population levels that ultimately can be related to crop-loss figures.

3.2 Assessment of levels of infestation

Since it is impracticable to count all the insects present in the field the pest population must be estimated from samples. Pests are usually sampled so that their abundance can be predicted, the losses attributable to them measured, and their damage prevented (Strickland, 1961). The critical importance of sampling as a basis for control decision has been reviewed by Gonzalez (1971) and the place of a sample survey in crop-loss estimation has been discussed by Church (1973). The method of sampling will vary according to the particular situation, e.g. whether the pest is in the soil or on the plant. In the latter case it must be decided which parts of the plant are to be sampled. If direct sampling of the pest population is impracticable, indirect estimates of the pest population may be possible from activities of the pest that are related to its number, such as feeding injury or number of faecal pellets. For example, in Tanzania assessment of attack by the stem borer, *Busseola fusca*, is by the simple method of counting the number of plants showing active larval damage to the leaves (Walker, 1960).

Methods of sampling and measuring populations range from trapping and marking to counting caterpillar frass. Sampling procedures must be adapted to suit different situations. Methodology is crucial to all ecological studies. The procedures employed determine the time invested by the investigator and in the end decide the validity and success of the results. To be of value, the method of sampling must be efficient, cheap and practical (see, for example, Beeden's (1972) use of peg board to estimate the pest status of cotton bollworms in Malawi). For a full discussion on sampling procedures, standard works (Morris, 1960; Kevan, 1962; Taylor, 1962; Pradhan, 1964; USDA, 1969; Nishida and Torii, 1970; FAO, 1971–73; CIBA-GEIGY, 1975; IRRI, 1976; Onsager, 1976; Khosla, 1977; Southwood, 1978; Sterling, 1979, etc.) should be consulted and the advice of an agricultural statistician sought before making insect counts. Nevertheless it is useful to bear in mind the following points:

Pest dispersion Initially information should be obtained on the distribution of the pest – whether it tends to aggregate, or is restricted to certain parts of the field, or if it is regularly distributed.

Considerable variations in pest density between different areas of the same field are common; sometimes the edges of a field may be more heavily infested. The density of a crop is also known to affect the distribution of the pest. Field boundaries where insects can find shelter also affect their distribution. Bardner & Fletcher (1974) in a review of distribution of insect infestations note that pest distribution can also be affected by any marked heterogeneity of the crop. Thus Harris (1962) and Walker (1960) found that stem borers of millet and maize prefer the largest plants in these crops. The time of attack can also greatly affect the distribution of insects on the plant (see Bardner & Fletcher, 1974 for details).

Number of samples Generally, the larger the number of samples the more accurate the population estimate. But the number of samples required and their size should be determined prior to sampling. Usually for larger populations, fewer samples are required and for small, densely distributed pests, many small samples are best. However, there is no *a priori* way of determining the number of samples and sample size. For each crop-pest combination, initial sampling must be done arbitrarily and the results statistically analysed for variation. Then the number of samples required to obtain an acceptable degree of variation can be estimated.

Sequential sampling, in which the number of samples required to estimate population density is determined as sampling progresses, was developed during World War II. Because of its value, sequential sampling was classified as restricted and was not made available for wider use until after World War II (Wald 1945, 1947; Statistical Research Group, 1946). When used in insect sampling, its main value is efficiency in that very few observations are needed when populations are very low or very high, and extensive sampling is required only at intermediate density levels (Harcourt, 1966). A sequential sampling system for bollworm eggs and larvae has been devised for cotton in Botswana (Ingram & Green, 1972) where, due to the low rainfall, the recommended insecticide carbaryl exerts control over both *Diparopsis* and *Heliothis* pests. Considerable time can be saved using such a system, particularly where infestation levels are very high or low. Bibliography of sequential sampling plans for insects is provided by Pieters (1978).

Sample bias Reliable estimates of pest incidence should be based on representative samples. As far as possible the sampling method should be free from bias. Not only sites but fields, units, etc. should be chosen using a table of random numbers and as far as possible average sampling conditions should be the same. Sometimes, however, it may be unrealistic to insist on a sampling method completely free from bias particularly if it adds greatly to the survey costs.

Sampling frequency Samples should be taken at regular intervals, e.g. once a week, fortnight, or month at fixed timings. Sometimes one sampling is all that is needed, e.g. sampling of grape colaspis, *Colaspis flavida* (Rolston & Rouse, 1960). In other cases regular sampling is required, e.g. the boll weevil, etc. on cotton. With some pests which complete only one or two generations per year, it is sometimes possible to obtain the required data by sampling for a particular stage only a few times each year.

Duplicate sampling It is desirable to use more than one sampling technique in assessing pest numbers if only to determine the relative efficiency of the methods employed. It is desirable initially to try various sampling techniques and even to use different techniques at various stages of plant growth.

Perennial sampling Neither the pest population nor the yield losses are stationary; they will change from year to year in a given locality. Experiments need to be conducted over a period of at least three years and the information should be up-to-date. This is particularly important when the cultural practices which affect the crop yield change from time to time. In many countries such as the United Kingdom assessment of pest importance is done by observations made all over the country, for several years. Routine estimates of the pest densities and losses due to them are made at fixed sites throughout the country. Such studies help to provide valuable information on regional distribution of the organisms as well as their pest status.

The importance of correct pest assessment in economic entomology cannot be over-emphasized. Such information is also invaluable in pest surveys, to assess pesticide trials, to study the biology and life cycle of pests, to study crop losses, and to use in forecasting pest outbreaks. The pest management decisions are usually based on one or more of these inputs. New computer and calculator technology have revolutionized ecology. New analyses and new presentations of the data are possible and sometimes it may indeed be possible to make good data out of bad. However, as noted by Southwood (1978), the sophisticated technology cannot 'make a "silk purse" of sound insight, out of a "sow's ear" of unreliable raw data or confused analytical procedure'.

3.3 Crop losses

Even when the damage to a crop appears heavy to the naked eye, the real losses of yield may be small and not necessitate control measures. However, there is widespread inability among growers and professional agriculturists to distinguish between damage and economic injury (Glass, 1975). Many a time the farmer spots infestation in the field, or is concerned about crop losses and is inclined to use chemical control measures to be on the safe side. The damage will depend on the stage of the crop. A young crop is usually more often damaged by pests. However, during the period of active growth the plant may be able to successfully withstand the pest attack by rapidly repairing damaged tissues and so exhibit little reduction in yield. Glass (1975) notes that it is difficult for many growers, insecticide salesmen and plant protection specialists to accept the well-documented fact that some levels of pest damage or the effect of rather dense population of some pest species has no measurable effect on yield or quality of the crop. Field experiments relating yield to stem-borer attack in Nigeria (Harris, 1962) and Ghana (Kumar, unpublished) indicate that loss of stand in maize (*Zea mays*) plots does not always result in loss of yield, and actually under certain conditions, loss of stands is compensated by the production of heavier cobs. Bardner (1968) observed similar phenomenon in his work on the effect of wheat bulb fly on the growth and yield of wheat. Experiments on guinea corn (Harris, 1962, 1964) showed that the yield per bored stem was higher than that per unbored stem. Actually it is now known that excellent yields of guinea corn can be obtained in the presence of high population of stem-borers (Ingram, 1958). Beneficial effects of pests on crop yield are discussed by Bardner & Fletcher (1974). Harris (1974) has suggested that increases of plant yield following insect damage may result from the early removal of apical dominance of plants growing with little competition for nutrients and to hormone-like effects produced by some sucking insects. He believes that the attainment of maximum crop yield may indeed sometimes require a certain density of 'pest' insects.

On the other hand some pests attack the crop most when it is fruiting. It is also not uncommon to find a succession of pests on a crop from vegetative to fruiting stage. Obviously it is best to record damage at suitable intervals during crop growth. Great variability in the effect of factors causing economic loss has rightly been emphasized by Judenko (1972). For example, in Nigeria, as mentioned earlier, according to Anonymous (1974), experimental defoliation of cassava plants after 9–11 months had no effect on the yield but at 7 months, and especially if there was repeated defoliation, the mean yield was reduced to about 60% of that of the control plants. Work of Page *et al*. (1980) suggests that defoliation of cassava plants by the grasshopper *Zonocerus variegatus* during the dry season in Nigeria causes no significant crop loss when damage occurs before leaf regeneration at the beginning of the following wet season. According to Pathak & Dyck (1972) stem-borer infestation of up to 10% deadheart in plants up to seven weeks after transplanting can be tolerated if the crop is later protected from stem-borers.

Low pest populations of insects may cause loss of crop beyond all proportions. Thus mealy-bug populations are very low in cocoa farms but they are capable of transmitting swollen shoot disease virus to a large number of cocoa trees which are ultimately killed. A population of ten crickets per square metre could cause very serious loss on a germinating cotton crop, whereas one five times as dense might not cause noticeable injury to full-grown cotton (Ahmad, 1954).

3.3.1 Reasons for assessing crop losses

As noted by Judenko (1972, 1973) different authors have advanced various reasons for studying crop losses and these include the following:

(1) The establishment of the economic status of specific pests; (2) to determine pest infestation intensity at which control measures need to be applied; (3) to assess the extent to which expenditure for intended control measures is justified, i.e. the economics of control measures; (4) to estimate the effectiveness of control measures; (5) to measure the effects of environmental factors on the loss of yield caused by pest attack;

(6) to provide information to pesticide operators to enable them to decide on the action to be taken to control the pests; (7) to assess the use of public funds to study pests; and (8) to give a basis for directing future research and agriculture planning.

Pinstrup-Andersen *et al*. (1976) emphasize that a knowledge of the relative importance of yield-limiting factors assists in establishing effective priorities in channelling research and extension resources and correcting the cause of crop loss. Actually, a knowledge of crop losses helps to focus attention on the harnessing of preventive measures by adoption of good agronomic practices such as growing of resistant varieties, use of cultural control methodologies and the development of forecasting of pest population trends with a view to timing the applications correctly, thereby achieving better control with less use of pesticides.

3.3.2 Methods of assessing crop losses

Walker (1977) discusses relation between infestation and yield under six categories (Fig. 2). The exact form of relation will of course depend on the time of attack, stage of pest or crop, and growing conditions.

Replicated field trials Entomologists have used replicated field trials to assess crop losses as a result of pest infestations, using randomized blocks or on randomly selected plots in fields. Here some plots are kept free from pests by blanket insecticide treatments or other control measures whereas the plants of the other plot are allowed to be damaged by naturally-occurring populations of the same pests. After conducting a number of experiments in an area and accumulating extensive data on pest intensity, their effect on yield reduction can be established. A plotting of points for each level of pest intensity and corresponding yield reduction will indicate the nature of mathematical regression. Using the results of a series of replicated field trials carried out at different locations and in different seasons in any one area, it is possible to calculate mean regression line for that area. Subsequently, just by estimating

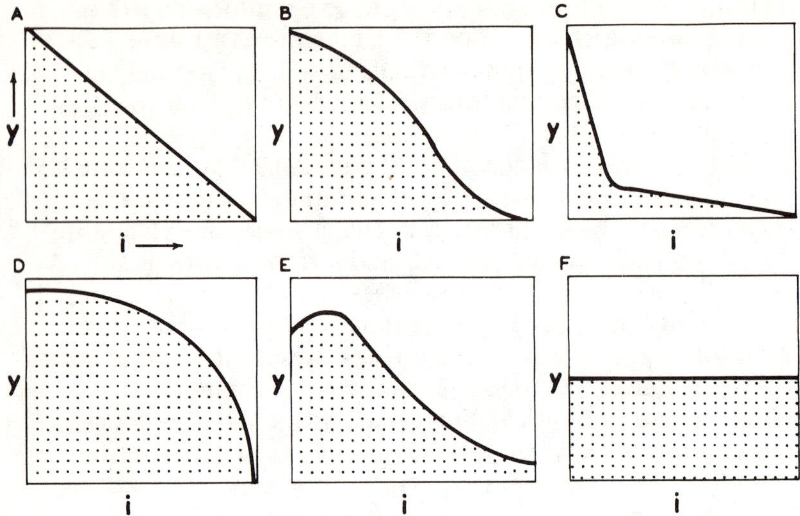

Figure 2 Examples of relationships between pest infestation and loss of yield (after Walker, 1977): A – loss of yield is directly proportional to infestation increase; B – relationship is sigmoid with many possibilities; C – a little increase in a very low population causes rapid loss of yield, but there is little further loss with increasing infestation; D – yield is related to the logarithm of the number of pests; E – yield rises with light pest attack; F – there is no change in yield with increasing pest attack (i = infestation; y = yield).

average reduction in yield for that particular area can be calculated. It has been assumed that the relationship of yield reduction to pest intensity is linear but the relationship is often logarithmic (Snedecor, 1956) or may be sigmoid in nature. For fully evaluating this type of relationship, appropriate statistical treatment is required.

Several authors have stated that it is dangerous to use an insecticide to assess the increase in crop production that resulted from the control of any given pest, unless the cumulative effect of the insecticide was ascertained, as it might directly or indirectly influence the status of many other insect species. Pickett (1954) commenting on the use of this method in assessing the value of sprays in the control of apple insects in Nova Scotia stated that the 'estimations were not worth the paper on which they were printed'. According to him the figures had no validity, because the dynamic nature of the environment was not taken into consideration. Reed (1972) points out that the 'unsprayed control' plots which many entomologists include within trials of insecticide treatments on cotton may give misleading information. The pest attack in such plots may differ from that which would occur on unsprayed cotton without adjacent, sprayed cotton. He concludes that unsprayed controls within many spraying trials serve no useful purpose. Judenko (1972) states that assessment of the actual yield and economic losses in plots treated with pesticide in the same experiment would only measure the effectiveness of a given level cf infestation. Despite the disadvantages mentioned above, replicated field trials have given very good results in crop-loss assessment studies. Indeed this method has been regularly used by workers studying crop losses due to attack by pests such as stem-borers (e.g. Harris, 1962; Chatterji *et al.*, 1969; Jotwani *et al.*, 1971; Rai *et al.*, 1978a, b, c, etc.) as well as pod-borers (e.g. Raheja, 1976).

Analytical method/paired plant method This method used by many authors consists of a comparison of the yields from two sets of similar plants growing under identical conditions. One set is attacked by a specific pest, the second is left free of infestation. The main principles of this method (Judenko, 1972) are as follows:

Let AY = actual yield per unit area (e.g. green or wet weight)
a = mean yield per uninfested or unattacked plant
b = mean yield per infested or attacked plant
p = percentage of plants infested or attacked
Then C = the coefficient of harmfulness can be calculated as follows:

$$C = \frac{(a - b)100}{a}$$

The percentage of economic loss (L) is

$$L = \frac{Cp}{100}$$

and the expected yield (W) in the absence of pest attack is

$$W = \frac{100\,(AY)}{100 - L}$$

The economic loss (LOS) is then given by

$$LOS = W - AY$$

Judenko (1973) discusses the use of analytical method for wheat, rye, barley, oats, rice, maize, sorghum and millet in cases when there is no compensatory yield of attacked plants as well as when compensatory yield of unattacked plants occurs. For full details of the procedures and formulae employed including methods of labelling individual plants for such studies Judenko's (1973) paper should be consulted.

Dinther (1971) uses the following formula for assessing rice yield losses caused by the stem borers *Rupela albinella* and *Diatraea saccharalis* in Surinam:

$$\text{Loss} = (a - b)np$$

Where loss is in kg ha^{-1}:

a = mean panicle weight of uninfested stem
b = mean panicle weight of infested stem
n = number of panicles per m^2 (or per ha)
p = percentage infestation

A number of other methods such as caging the crop or plants to keep pests out, or removing pests artificially, or infesting plants with artificial infestation have been used by investigators to study the relation between infestation and yield (e.g. see Gangrade *et al.*, 1975 and others). Obviously, here an understanding of the crop physiology in relation to growth and yield of the plant as well as the influence of abiotic factors on them is essential. Reliable figures on regional and national losses would ultimately form the bases of good pest management programmes.

Recently Pinstrup-Andersen *et al.* (1976) have suggested another procedure for estimating yield and production losses in crops. The method consists of the following steps:

(1) Selection of a representative panel of farms where data collection would take place at predetermined intervals.

(2) Estimation of yield losses caused by factors such as

insects, diseases, weeds, soil characteristics and fertility etc. Yield reduced by different causes can be shown diagrammatically in a crop-loss profile (Fig. 3). Yield losses are determined on the basis of a production function analysis. Each regression coefficient is multiplied by the average value of the particular yield-limiting factor. This provides an estimate of the overall impact of this factor on sampled yield.

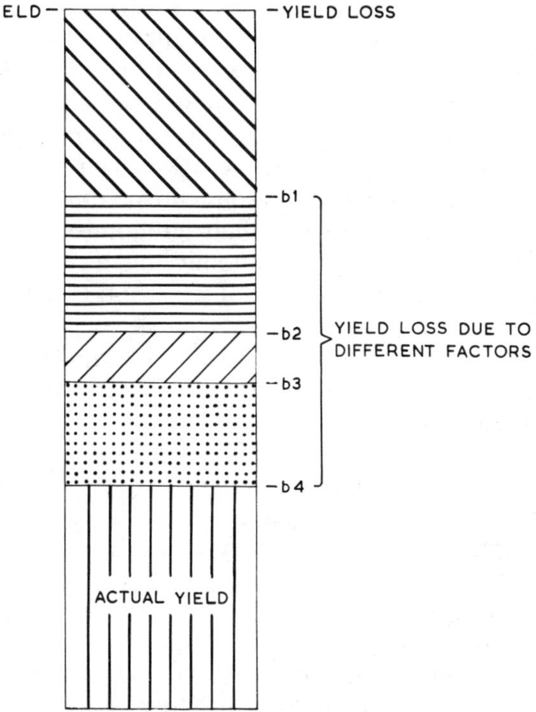

Figure 3 Crop loss profile (diagrammatic) showing yield losses due to different factors (based on a paper by Pinstrup-Andersen *et al.*, 1976).

(3) Market value of losses is assessed by estimating expected price changes due to the production increases which would occur, if the factors mentioned above were removed. Thus actual benefits that would result from the use of particular management policies can be determined and a decision made whether or not to use such control measure.

However, it should be mentioned that the method of Pinstrup-Andersen *et al.* (vide supra) is rather impractical considering the problems encountered in estimating the impact of one pest even without introducing weeds, diseases, etc.

3.4 Use of cost/potential benefit ratio

In deciding to adopt control measures, Ordish (1952) has suggested that it is useful to know the relationship between cost of control and potential benefit in terms of increased yield. This is an ideal that should be aimed at in pest management programmes. However, information such as net profit per unit area as a result of a particular treatment is known in only a few instances. For such a knowledge, as noted by Toms (1967), a multidisciplinary team consisting of an agronomist, an entomologist or mycologist, an economist and a computer will be required to work over a number of years and this itself is likely to be uneconomical.

In practice, the economics of control is determined in terms of an average yield of so many lb or kg per acre or hectare that can be expected by an efficient control of the main pest. The question of whether usage of insecticides is economical in particular situations has largely remained unanswered. It is quite possible that in many cases, to get any harvestable crop, uneconomic spraying must be employed. While in other cases the pest incidence may have become quite low or negligible but the farmer nevertheless continues to use unnecessarily high dosages of the pesticide.

Practical experience has shown that most farmers have a pretty good idea of how much they produced and what they got for their product. They have expectations for yield and price, although these are not always realized. Often yield and price vary to help maintain an equilibrium, i.e. a poor year for production is compensated by higher prices and *vice versa*. Finally, there are many instances where an insecticide or schedule of insecticides are applied as an insurance. Most fruits, several vegetables and some field crops affected by soil insects are so treated. This is done because past experience demonstrates that on the average such a procedure is profitable. Thus the concept of tailoring treatments to the currently observed pest population is not applicable in all cases. Another case where the knowledge of economic threshold is practically valueless is where the cost of estimating the pest population is greater than the cost of treatment (see, for example, Rolston & Rouse, 1965).

3.5 Relationship between pest intensity and crop loss

(1) *Determination of injury point* Earlier, the importance of reliable information on yield losses as a result of pest attack was emphasized. That point at which injury due to density of an insect population begins to reduce crop yield is difficult to determine and the following factors should be taken into consideration in arriving at this point:

(*a*) quantity and quality of the crop
(*b*) economic value of the crop
(*c*) cost of remedial measures

Hensley, Concience and graduate assistants carried out research on sugar-cane pests at Louisiana (U.S.A.) from 1965–1970. They found 5% stalk infestation as the economic threshold for the borer, *Diatraea saccharalis*. Uncontrolled infestation at this level was found to result in the following:

(*a*) about 10% internodes of the plants were bored by larvae during a crop season
(*b*) a yield reduction of 3.75 t of sugar-cane ha^{-1} valued at \$41.25 (\$27.50 t^{-1}) from the resultant infestation.

To control continuous, heavy, seasonal infestations, the grower needed to pay about \$32.50 ha^{-1} and this covered expenses in respect of the following:

(*a*) cost of survey and application services, and
(*b*) cost of insecticide (3 applications).

(2) *Procedure for determining levels of infestation* Insect counts are made at weekly intervals in each field by examining 50 plants, at least 1 m apart, at 6 locations in each 15 or 40 ha of sugar-cane. The number of stalks with small larvae in leaf sheaths is counted and the treatment is recommended only after meeting the following criteria:

(*a*) internodes are formed above the soil level, and
(*b*) infestation attains a 5% level.

Advantages of the survey method outlined above are as follows (Hensley, 1971):

(*a*) it permits management of the pest population that is injuring the developing stalks and precludes the fixed insecticides application schedules;

(*b*) it permits rapid detection of bad control due to any of the following reasons:

(*i*) bad timing of the spray schedules,

(*ii*) faulty application of insecticide,

(*iii*) weak formulation of the insecticide, and

(*iv*) insecticide resistant borer populations.

The survey method also helps to take into account the following variables:

(*a*) differences in susceptibility of varieties of sugar-cane to borer attack so that less insecticide needs to be used on resistant varieties;

(*b*) the role of biotic (e.g. natural enemies) and abiotic (e.g. weather) factors which affect actual insect counts in the field and on which the treatment recommendations are based.

Large firms employ professional entomologists who continuously monitor the incidence of infestation. Hensley (1971) estimates that one entomologist can adequately survey about 3000 ha per week. For farms of less than 80 ha, owners do their own monitoring. It is to be noted that the 5% economic injury level given above is not applicable to all other crops and must be worked out for each situation bearing in mind the criteria mentioned earlier.

3.6 Examples from Africa

Detailed studies, such as the examples given above, are rare in the tropics and subtropics. However, information on relationships between pest intensity and crop losses and procedures to determine such relationships are available for the following crops and pests in Africa (FAO, 1971–73).

Cotton and cotton jassids in Sudan
Cotton and cotton whitefly in Sudan
Maize and other stem-borers in Nigeria and Tanzania
(see Walker, 1971 for relationship between pest infestation of sugar-cane by *Eldana saccharina* and yield of sugar in Tanzania).

Sorghum and sorghum midge in Nigeria and Ghana
Sorghum and covered smut in Nigeria
Haricot bean and bean-fly in Tanzania
Maize and snout beetle complex in Zimbabwe
Maize and termites in Tanzania
Cowpea and coreid bugs in Nigeria (see Raheja, 1976 for an assessment of losses caused by insect pests to cowpeas in Northern Nigeria)
Cowpea and African pea moth in Nigeria
Rice and African rice-borer in Madagascar
Coffee and coffee bug in Kenya, Uganda and Tanzania
Sugar-cane and white grubs in Mauritius.

3.6.1 Economic thresholds and peasant agriculture

The question of the applicability of a single economic threshold to peasant farms has received very little attention. As noted by Farrington (1977), assessment of pest densities and of the pest/host plant relationship as well as costs of control applications are likely to vary widely among farms and from season to season. It should be realised that in peasant agriculture such as in most parts of Africa, farms are usually very small and crops in them frequently occur at different stages of growth within a small area. Therefore, as observed by Farrington (1977), the use of a single threshold and pest count is likely to result in an 'over-investment in pest control for some farmers, under-investment for others and the economically optimal investment for perhaps only a minority'. Therefore, there is a pressing need for a flexible system of threshold such as is used in the control of cotton bollworms in Malawi (Beeden, 1972). Here a scouting aid 'pegboard' consisting of a wooden block with three rows of holes is used. A peg is moved along each row to record the number of plants sampled as well as egg counts for two main bollworm pests. The decision to spray is taken as soon as the egg count crosses the appropriate threshold line marked on the board. This device assists the farmers in making their own insecticide recommendation, and involves no reading or writing.

3.7 Summary

To develop and utilize effective pest control strategies,

information is required on crop-yield reduction relative to pest density. The determination of an economic threshold of an insect pest on a crop is dependent on measuring infestation levels and the degree to which each level affects the harvested crop. To obtain this information basic statistical requirements must be met. An accurate and efficient sampling technique for assessing pest infestation, adequate sample size, etc. should be used. Finally, incidence levels are related to yield losses. In short, crop protection economics involves much more than simple cost accounting. Despite some examples of pest intensity/crop loss available, the need for determining accurately the economic thresholds for pest management purposes is nowhere greater than in the developing parts of the world, especially the African continent where conditions of peasant agriculture are very different from those of the highly mechanized systems of the western world (Farrington, 1977). Such information is absolutely vital to the development of any meaningful control measures. Stern (1973) concludes 'lack of knowledge on the yield/pest density ratio has unquestionably led to frequent erroneous judgements and unnecessary control measures. There is sufficient evidence to indicate that pest control can be placed on a much higher level of competency in the form of applied population ecology'.

3.8 Literature cited

Ahmad, T. (1954). *Rep. 6th Commonw. ent. conf.* p. 87.

Anonymous (1974). Control of *Zonocerus variegatus* L. in Nigeria. Interim Report. Jan. 1972–July 1973. *Centre for Overseas Pest Research, London*, 1–17.

Bardner, R. (1968). Wheat bulb fly, *Leptohylemia coarctata* Fall., and its effect on the growth and yield of wheat. *Ann. Appl. Biol.*, **61**, 1–11.

Bardner, R. & Fletcher, K. E. (1974). Insect infestation and their effects on growth and yield of field crops: a review. *Bull. ent. Res.*, **64**, 141–60.

Beeden, P. (1972). The pegboard – an aid to cotton pest scouting. *PANS,* **18(1)**, 43–5.

Carlson, G. A. (1973). Economic aspects of crop loss control at the farm level. In *Crop Assessment Methods*. F.A.O. manual on the evaluation and prevention of losses by pests, diseases and weeds (not sequentially paginated).

Chatterji, S. W., Young, W. R., Sharma, G. C., Sayi, I. V., Chahal, B. S., Khare, B. P., Rathore, Y. S., Panwar, V. P. S. & Siddiqui, K. H. (1969). Estimation of loss in yield of maize due to insect pests with special reference to borers. *Indian J. Ent.*, **31(2)**, 109–15.

Chiang, H. C. (1979). A general model of the economic threshold level of pest populations. *F.A.O. Plant Protection Bulletin*, **27(3)**: 71–3.

Church, B. M. (1973). The place of sample survey in crop loss estimation. In *Crop Loss Assessment Methods*. F.A.O. manual on the evaluation of losses by pests, diseases and weeds (not sequentially paginated).

CIBA-GEIGY (1975). Field trials handbook (loose leaf).

Dinther, J. van (1971). A method of assessing rice yield losses caused by the stem borers *Rupela albinella* and *Diatraea saccharalis* in Surinam and the aspect of economic thresholds. *Entomophaga*, **16(2)**, 185–91.

FAO (1971–73). Crop loss assessment methods: F.A.O. manual on the evaluation of losses by pests, diseases and weeds. F.A.O. and Commonwealth Agricultural Bureaux, Farnham Royal, England (1977 Supplement, not sequentially paginated).

Farrington, J. (1977). Economic threshold of insect pest infestation in peasant agriculture: a question of applicability. *PANS*, **23(2)**, 143–8.

Gangrade, G. A., Singh, O. P. & Matkar, S. M. (1975). Soybean yield losses in response to damage by varying levels of three lepidopterous larvae. *Indian J. Ent.*, **37(3)**, 225–9.

Glass, E. H. (1975). Integrated pest management: rationale, potential, needs and implementation. *E.S.A. Special Publication*, 75–2, 1–141.

Gonzalez, D. (1971). Sampling as a basis of pest management strategies. *Proc. Tall Timbers Conf. Ecol. Anim. Control Habitat Management.*, **2**, 83–101.

Harcourt, D. G. (1966). Sequential sampling for use in control of the cabbage looper on cauliflower. *J. Econ. Ent.*, **59**, 1190–2.

Harris, K. M. (1962). Lepidopterous stem borers of cereals in Nigeria. *Bull. Ent. Res.*, **53(1)**, 139–71.

Harris, K. M. (1964). The effect of lepidopterous stem borers on the yields of cereals in N. Nigeria. *Proc. 12th International Congress of Entomology*. p. 610.

Harris, P. (1974). A possible explanation of plant yield increase following insect damage. *Agro-ecosystems*, **1**, 219–25.

Hensley, S. D. (1971). Management of sugarcane borer populations in Louisiana, a decade of change. *Entomophaga*, **16(1)**, 133–46.

Ingram, W. R. (1958). The lepidopterous stalk borers associated with Gramineae in Uganda. *Bull. Ent. Res.*, **49**, 367–83.

Ingram, W. R. & Green, S. M. (1972). Sequential sampling for bollworms on raingrown cotton in Botswana. *Cotton Grow. Rev.*, **49(3)**, 265–75.

IRRI (1976). Standard evaluation system for rice. International Rice Research Institute, Los Baños, Philippines. 64 pp.

Jotwani, M. G., Chandra, D., Young, W. R., Sukhani, T. R. & Saxena, P. N. (1971). Estimation of avoidable losses caused by the insect complex on Sorghum hybrid CSH I and percentage increase in yield over untreated control. *Indian J. Ent*., **33(4)**, 375–83.

Judenko, E. (1972). An assessment of economic losses in yield of annual crops caused by pests, and problem of economic threshold. *PANS*, **18(2)**, 186–91.

Judenko, E. (1973). Analytical method for assessing yield losses caused by pests on cereal crops with and without pesticide. *Tropical Pest Bulletin*, **2** (1–29 pp), *Centre For Overseas Pest Research, London.*

Kevan, D. K. M. (1962). *Soil Animals*. Witherby, London. 237 pp.

Khosla, R. K. (1977). Techniques for assessment of losses due to pests and diseases of rice. *Indian J. Agric. Sci.*, **47(4)**, 171–4.

Morris, R. F. (1960). Sampling insect populations. *Ann. Rev. Ent*., **5**, 243–64.

Nishida, T. & Torii, T. (1970). Handbook of field methods for research on rice stem borers and their natural enemies. *International Biological Programme Handbook* 14, Blackwell, Oxford. 132 pp.

Onsager, J. A. (1976). The rationale of sequential sampling, with emphasis on its use in pest management. *Agricultural Research Service, United States Department of Agriculture, Hyattsville, Maryland*. Technical Bulletin No. 1526, 1–19.

Ordish, G. (1952). *Untaken Harvest.* Constable, London. 171 pp.

Page, M. W., Harris, J. R. W. & Youdeowei, A. (1980). Defoliation and consequent crop loss in cassava caused by the grasshopper *Zonocerus variegatus* (L.) (Orthoptera: Pyrgomorphidae) in southern Nigeria. *Bull. ent. Res*., **70**, 151–63.

Pathak, M. D. & Dyck, V. A. (1972). Rice insect control studies at IRRI. Workshop On Integrated Rice Pest Control in Southeast Asia. 9–12 May 1972. IRRI. Los Baños, Philippines. 16 pp.

Pickett, A. D. (1954). *Rep. 6th Commonw. ent. Conf.* p. 87.

Pieters, E. (1978). Bibliography of sequential sampling plans for insects. *Bull. Ent. Soc. Amer*., **24(3)**, 372–4.

Pinstrup-Andersen, P., de Londono, N. & Infante, M. (1976). A suggested procedure for estimating yield and production losses in crops. *PANS*, **22(3)**, 359–65.

Pradhan, S. (1964). Assessment of losses caused by insect pests of crops and estimation of insect population. In Pant, N. C. (Ed.) *Entomology in India, Indian J. Ent. Silver Jubilee Number. New Delhi, Entomological Society of India*, pp. 17–58.

Raheja, A. K. (1976). Assessment of losses caused by insect pests to cowpeas in Northern Nigeria. *PANS,* **22(2)**, 229–33.

Rai, S., Jotwani, M. G. & Jha, D. (1978a). Methodology for estimating shootfly damage and grain yield relationship in Sorghum. *Indian J. Ent.,* **40(2)**, 121–5.

Rai, S., Jotwani, M. G. & Jha, D. (1978b). Estimation of losses at different levels of fly infestation in Sorghum. *Indian J. Ent.*, **40(3)**, 254–60.

Rai, S., Jotwani, M. G. & Jha, D. (1978c). Economic injury level of shootfly, *Atherigona soccata* (Rondani) on Sorghum. *Indian J. Ent.*, **40(2)**, 126–33.

Reed, W. (1972). Uses and abuses of unsprayed controls in spraying trials. *Cott. Gr. Rev.*, **49**, 67–72.

Rolston, L. H. & Rouse, P. (1960). Control of grape colaspis and rice water weevil by seed or soil treatment. *Agricultural Experiment Station, Division of Agriculture, University of Arkansas, Fayettville. Bulletin,* **624**, 3–10.

Rolston, L. H. & Rouse, P.(1965). The biology and ecology of the grape colaspis, *Colaspis flavida* Say, in relation to rice production in the Arkansas Grand Prairie. *Agricultural Experiment Station, Division of Agriculture, University of Arkansas, Fayettville, Bulletin,* **694**, 31 pp.

Snedecor, C. W. (1956). *Statistical Methods*. 5th Ed. Iowa State College Press, U.S.A. 485 pp.

Southwood, T. R. E. (1978) *Ecological Methods with Particular Reference to the Study of Insect Populations*. 2nd Ed. Chapman and Hall, London, 524 pp.

Statistical Research Group, Columbia University (1946). Sequential analysis of statistical data: Applications. Columbia Univ. Stat. Res. GP. 255 (rev.) and *Office Sci. Res. Dev. App. Math. Panel* Rep. 30 2R. (rev.) Columbia Univ. Press. N.Y. 393 pp.

Sterling, W. L. (ed.) (1979). Economic thresholds and sampling of *Heliothis* species on cotton, corn, soybeans and other host plants. *Southern Cooperative Series Bulletin 231. Texas.* 159 pp.

Stern, V. M. (1973). Economic thresholds. *Ann. Rev. Ent.,* **18**, 259–80.

Stern, V. M., Smith, R. F., van den Bosch & Hagen, K. S. (1959). The integration of chemical and biological control of the spotted alfalfa aphid. Part 1. The integrated control concept. *Hilgardia,* **29**, 81–101.

Strickland, A. H. (1961). Sampling crop pests and their hosts. *Ann. Rev. Ent.*, **6**, 201–20.

Talpaz, H. & Frisbie, R. E. (1975). An advanced method for economic threshold determination: a positive approach. *Southern Journal of Agricultural Economics*, December, 19–25.

Taylor, R. L. (1962). Suction methods for sampling arthropods at and above ground level. *Progress in Soil Zoology*, **1**, 217–21.

Toms, A. M. (1967). Some aspects of the economics of crop protection. *PANS*, **13**, 135–42.

USDA (1969). Survey methods for some economic insects. Agricultural Research Service, ARS 81–31, Hyattsville, U.S.A. 137 pp.

Wald, A. (1945). Sequential tests of statistical hypotheses. *Ann. Math. Stat.*, **16**, 117–86.

Wald, A. (1947). *Sequential Analysis*. John Wiley and Sons, New York. 212 pp.

Walker, P. T. (1960). The relation between infestation by the maize stalk borer, *Busseola fusca* and yield of maize. *Ann. Appl. Biol.*, **48(4)**, 780–6.

Walker, P. T. (1971). The relation between infestation of sugarcane by a stemborer, *Eldana saccharina* Walk. (Lep., Crambidae) and yield of sugar in Tanzania. *Proceedings 13th International Congress of Entomology, Moscow*, **2**, 413–14.

Walker, P. T. (1977). Crop losses: some relationships between yield and infestation. *Med. Fac. Landbouww. Rijksuniv. Gent*., **42(2)**, 919–26.

4 Pest Forecasting and Predictive Monitoring

4.1 Necessity of forecasting and monitoring of pest attacks

Insecticides have been and to a lesser extent still are used routinely as an insurance against losses due to pest attacks. However, their indiscriminate use has sometimes resulted in a number of important and undesired side effects and it has now become necessary to apply insecticides only when absolutely necessary. Since pesticides are likely to remain the principal weapon against pests in the foreseeable future, it is essential to use them rationally. To do so it is helpful to be able to detect and predict outbreaks. Such forecasts can provide responsible agencies as well as the farmers time to organize control measures. In this way control applications can be correctly timed resulting in better control and less use of pesticide. Forecasting has been especially necessary and successful with migratory locusts and the African armyworm.

4.2 Practice of forecasting and monitoring pest attacks

Forecasting of pest attack and the best time at which to treat the infestation depends on establishing relationship between:

(*i*) stage of development of the crop,
(*ii*) stage of development of the pest, and
(*iii*) environmental factors.

On the basis of pest correlation, future predictions can be made for a few pests. For successful forecasting, a good knowledge of the density of infestation and its effect on the crop is essential. For this purpose, single, standard, quantitative, accurate, repeatable, economic and meaningful methods of assessing pest levels are required. This enables us to estimate probable crop loss in relation to pest incidence. The expected crop loss can then be costed in relation to the investment required for the application of control measures and financially sound decisions taken. A thorough knowledge of the biology and distribution of pests is necessary in order to have a simple monitoring system at the source of infestation, during dispersal from the crop and on the crop itself.

In practice, the forecasting of pest attacks has seldom been a simple operation. Insect populations, as well as climatic parameters, are subject to unpredictable fluctuations and these can upset well established prediction techniques which may be further affected by small changes in agricultural practices. The importance of a knowledge of the dynamics of the pests was stressed earlier and many variables must be taken into account in making forecasts. Some of the factors – such as cost of control operations – can be estimated fairly accurately but others – for example, the yield expected from interaction between pest, plant and environment – can only be forecast within wide limits. In addition, different species provide particular forecasting problems which may require different approaches to their solution. Further, especially in the case of a migratory pest a network of recording staff and offices is necessary to collect data and transmit them quickly and regularly to a central office where the information will be processed to make forecasts. Finally it will be necessary to disseminate the forecast information to users (e.g. farmers) by radio, television, newspaper or post.

4.3 Examples of pest forecasts

4.3.1 Pests in soil

Organisms in the soil, including immature stages of some flying insects, may be considered virtually 'immobile' and hence their population fluctuations can be predicted with some certainty (see Way & Cammell, 1973 for references). The relatively stable soil climate helps in making usually reliable forecasts of pests and diseases. Success has been reported in forecasting the incidence of some nematodes such as eel-worms (*Heterodera* spp.) and Collembola. Such information has been found useful in making decisions on long term crop rotations.

4.3.2 Air-borne pests

Forecasts based on weather conditions Environmental factors, such as temperature and rainfall, have been used in forecasting incidence of pest attack (see Edwards & Heath, 1964 for a comprehensive summary) and it appears that careful work in this direction would pay dividends. Some variables commonly monitored are: rainfall, relative humidity, solar radiation, ambient air temperature, soil temperatures at different depths, temperature within crop canopy, wind speed and wind direction. In Africa, outbreaks of the red locust (*Nomadacris septemfasciata*) have been forecast from an index of the previous year's rainfall. Symmons (1959) found that a very high level of Lake Rukwa is correlated with small numbers of the red locust. When the lake level is lower, the size of the adult locust population is negatively correlated with the total rainfall of the last but one wet season and positively correlated with the size of the preceding parental population. A multiple regression, using the following equation, was calculated:

$$Y = 6.518 - \frac{(0.160)}{25.4} x_1 + 0.425 x_2 + \frac{(0.092)}{25.4} x_3$$

where Y = adult population in mid-dry season,

x_1 = mean of the rainfall, in mm, of the Rukwa Valley and drainage basin during the last but one wet season,

x_2 = the infestation level of the breeding population,

x_3 = the total rainfall of the Rukwa Valley, in mm, in preceding October–December.

Lea (1958) in South Africa found that changes in the numbers of the brown locust (*Locustana paradalina*) are linked with the previous early summer rainfall, a dry early summer being followed by an increase and a wet early summer by a decrease in locust numbers in the following year. Smith (1954) found that outbreaks of grasshoppers in Kansas have followed two years in succession of subnormal rainfall and that low numbers have occurred when rainfall has been above normal in a given summer and the year before. In the drier northern parts of the Gezira (Sudan), Joyce (1959) established that poor pre-sowing rains resulted in increased jassid damage to cotton. It has been shown that poor rain caused poor growth of the cotton plants which are unable to utilize all the nitrogen available in the plant leaves. Thus the nitrogen level in years of poor pre-sowing rains is much higher than in normal years and is a known cause of increased jassid infestation (Joyce, 1953). The regression of the mean concentration of nitrogen in the leaf material of cotton plant on pre-sowing rains can be expressed by the following equation:

$$X_2 = 4.07 - 0.0031\, x_1$$

where X_2 = mean concentration on leaf nitrogen during September–October

x_1 = rainfall in mm during July–August.

Thus in years of poor pre-sowing rains the total quantity of nitrogenous fertilizer applied must be reduced. But recommendations of fertilizer application would have to be revised if cultural practices such as the cultivation of legumes were done in rotation with the cotton crop. Thus outbreaks of economic significance can be successfully predicted.

Joyce (1959) has shown that at least half of the deleterious effect of poor pre-sowing rains on Gezira yields is the pest attack which can be eliminated by insecticidal sprays. Thus by such predictions insecticides can be applied selectively to those areas of the Sudan Gezira where their effect is likely to be most profitable.

Forecasts based on outbreak areas Outbreak areas are ecologically suitable localities where the insect has the greatest possibility of developing and therefore of making an outbreak. These areas not only possess suitable environmental conditions such as rainfall, temperature, humidity, etc. but favourable vegetation features as well. The study of the optimum ecological conditions permits a definition in terms of time and space of the areas referred above.

Forecasts of locust infestation Considerable work has been done on locust forecasting. The desert locust (*Schistocerca gregaria*)

is capable of breeding over very extensive areas after a plague and has a very wide range of distribution and invasion. Effective control required warnings of both nymphs and adults and their potential economic threat. Currently, the responsibility of desert locust (*Schistocerca*) forecasting is handled by some five regional organizations co-ordinated by FAO (see Betts, 1976 on the forecasting of infestation of this locust). The service forecasts occurrence and movement of swarms from the present state of locusts and weather events, e.g. wind convergence, rainfall and temperature.

Locusta migratoria migratorioides occurrences are largely confined after a plague and will be used to illustrate the forecasting of locust infestations (see Alomenu, 1977 for details):

(1) The Middle Niger Delta in Mali, the Lake Chad Basin and possibly the Blue Nile and Kassala provinces of the Sudan are the areas ecologically most suitable for the multiplication of locusts' solitary populations, leading to phase transformation and consequent incipient hopper-band formation and swarm development.

(2) The flood plains and the surrounding semi-arid lands of the Mali outbreak area and Lake Basin provide excellent seasonal habitats which enable the various generations of *Locusta* to survive and breed continuously throughout the year.

(3) The insects move out of collapsing habitats such as drying vegetation or an area of very high soil moisture/relative humidity to colonize new habitats of optimal ecological conditions. Normally, the direction of these movements coincides with the predominant wind direction but the solitary *Locusta* can found new colonies against the general wind direction.

(4) The best way to control the locusts is to prevent plagues rather than to suppress them after they have begun. For this purpose permanent surveillance has been established to detect hopper bands and incipient swarms in their early stages and destroy them before any escape to surrounding countries becomes possible.

(5) Qualitative sampling of *Locusta* population is done by examining a randomized sample of adults, nymphs and sometimes egg-pods, and aims at establishing the age and the

state of development of the individuals and their phase status. The sampling may be done by foot counts, vehicle counts, aerial sampling by means of a helicopter or fixed-wing aircraft. These observations, coupled with the quantitative and allied estimates and as well as meteorological information permit a reliable assessment of the future *Locusta* situation.

(6) In the Mali outbreak area, *Locusta* concentration and gregarization take place during the period of retreating flood (November–April), when favourable habitats are restricted by flood, enforcing immigrant locusts from the Sahelian zones to concentrate and oviposit in limited areas. The preventive control strategy aims at awaiting the development of first and second flood recession generations in these areas and to attack them as the incipient hopper bands develop.

(7) The Mali outbreak area is divided into 16 scouting sectors where mobile scouting teams radio locust information directly to headquarters. A daily *Locusta* situation report is correlated with meteorological and ecological information which are analysed. This provides a forecasting system which allows for a proper planning of preventive control operations.

Forecasting system for African armyworm The African armyworm, *Spodoptera exempta* (Lepidoptera, Noctuidae) is one of the most serious pests of Graminae in East Africa and can suddenly appear in large numbers over vast areas and sometimes an entire crop or pasture may be lost. A warning service which encourages early discovery and reporting and enables the farmers and control agencies to organize operations in the outbreak areas has been in service for the last twelve years. The basic considerations for these forecast areas are as follows (see Odiyo, 1977 for details).

(1) Continuous assessment of changes in adult and larval populations, with special reference to weather, particularly wind-shifts and rainfall. Changes in moth populations are assessed by a network of some 37 Robinson-type traps from which nightly caught totals of male and female *S. exempta* are each week telegraphed to headquarters from Tanzania, Kenya and Uganda.

(2) Changes in populations of larvae are assessed by estimating the dates of potential moth emergence from known outbreak areas, while the relationship between moth catches and egg laying is assessed by estimating inferred egg laying dates of reported outbreaks of larvae. The two estimates thus show the distribution in time and space of major source areas.

(3) Wind and rainfall records from the Meteorological Department are then examined each week for clues of possible convergence centres where moths which move down-wind are likely to congregate.

(4) Comparison is made of the current catches with those of similar periods in past years over the same areas, with notes on full and new moon phases (full moon competes with traps), thus making allowance for high or low catches in traps.

(5) Using the data mentioned above forecasts of the possibility of outbreaks of larvae are made. They contain a statement in English and Swahili of the general distribution of moth and larvae, the stage of development of known outbreaks, and a statement of the probability of larvae occurring in new areas within the next week. They are posted every week to district and provincial agricultural offices throughout the region. A short version is also released to the press, radio and television networks in East Africa.

(6) All forecasts are critically assessed every year so that forecasters can identify any errors or omissions.

Some other examples of pest forecasting and monitoring From the foregoing account it is clear that pest forecasting and predictive monitoring has a major role to play in control of pests. The forecasts must, however, be based on a continuous assessment of population distribution, including both movement and breeding, and interpreted in conjunction with factors such as weather systems, which govern these developments. For this reason, in the United States of America, Europe and Japan where some very careful work has been done, forecasting of pest attacks has been made with reasonable success, at least in the case of some species.

United States Department of Agriculture, Animal and Plant

Health Inspection Service issues a weekly Plant Pest Report, based on detailed quantitative information. These are invaluable for forecasting purposes. There are local systems for forecasting maize, alfalfa, fruit and cotton pests, and some are highly computerized (see Haynes *et al*., 1973, Croft *et al*., 1976 and Fulton & Haynes, 1977). For codling moth (*Cydia pomonella*) pheromone traps are used to forecast egg laying and larval sampling and temperature summation is used to time spraying (Batiste *et al*., 1973). This results in a great reduction in sprays required for the control of the pest. A 20–100% saving in insecticide sprays has resulted from forecasting the time of emergence of the fruit-fly, *Rhagoletis indifferens* in cherry in Canada, using attractant traps (AliNiazee, 1976; 1978).

In the United Kingdom, the Ministry of Agriculture issues a weekly crop disease survey which is used to give detailed warning to farmers through local press, radio and television. The most important *pest and crop* situations are monitored from random field samples sometimes together with damage and pesticide usage. Computerization is in progress. There is also a nationwide suction and light trap survey used in forecasting aerial pests, and in issuing a weekly aphid bulletin (Taylor, 1973). Sugarbeet aphids are monitored from field samples and forecast from frost-days and April temperature (Hull 1968, Watson *et al*., 1975). Bean aphids are forecast from egg numbers on alternative winter hosts (Way *et al*., 1977, Gould & Graham, 1977). Fruit and pea moths are forecast from pheromone trap counts. Wheat bulb-fly (Bowden & Jones, 1979) is forecast from egg counts and from light trap catches.

In Japan, the Ministry of Agriculture issues national and local forecasts of pest outbreaks on rice, fruit and tea (Yasuo, 1971). Second generation stem-borer attack on rice by *Chilo suppressalis* is forecast from percentage hills attacked by first generation larvae, using a linear regression between the percentage of infested hills and the mean number of infested stems per hill. Account is also taken of the number of eggs laid and the parasitism rate (Torii, 1971). The larval density of *Spodoptera litura* at which to apply pesticide is forecast from the relation between yield and numbers of larvae in August which in turn is related to the number of eggs and the moths caught in pheromone traps.

Use of computers in pest forecasting As discussed above reliable forecasting involves monitoring many weather and crop condition variables. A very large amount of such information is required to develop the crop and pest population models. These models assist in predicting performance of pest populations and their impact on crops. Computer technology has been shown to be especially useful for monitoring weather variables over a wide range of environmental conditions. Computers are further useful in statistically analysing rapidly a large body of data as well as for organizing and executing complex simulation models. Some simple decision models are currently being executed on computers to assist farmers in certain areas in the United States to make management decisions based on economic feasibility. If the projected loss is greater than the cost of control tactics, then the decision model recommends that the control measures be implemented. If on the other hand the projected loss is less than the cost of the control measure, the decision model recommends that the situation be reassessed at a later date in the light of more available information.

4.4 Summary

The purpose of pest forecasting is to predict whether a pesticide will be needed and when it should be applied. All forecasting is based on relationships: between the event to be forecast, the critical stage of the crop or pest, and some related climatic or biological factor. Generally for successful forecasting, a good knowledge of the minimum pest density that justifies the cost of the control measures is necessary. For this purpose, continuous assessment of pest population, distribution, including both movement and breeding, and integration with factors such as weather systems which govern such developments, is required. But it must be recognized that different pest species provide particular forecasting problems which may require different approaches to their solution. In countries where very accurate research on pests has been done, forecasting of pests attacks has been made with reasonable success and this holds much hope for the world's agriculture. One may say with reasonable certainty that pest forecasting and predictive monitoring is the applied entomology of the future.

4.5 Literature cited

AliNiazee, M. T. (1976). Thermal unit requirements for determining adult emergence of the Western Cherry fruit fly (Diptera: Tephritidae) in the Willamette Valley of Oregon. *Environmental entomology*, **5(3)**, 397–402.

AliNiazee, M. T. (1978). The Western Cherry fruit fly, *Rhagoletis indifferens* (Diptera: Tephritidae) 3. Developing a management program by utilizing attractant traps as monitoring devices. *Canad. entomol.*, **110**, 1133–9.

Alomenu, H. S. (1977). The evolution of a new preventive control strategy in the African Migratory locust (*Locusta migratoria migratorioides* R. & F.) *Proc. 3rd Scientific Meeting Int. Centr. Ins. Physiol. Ecol.*, 1976, 34–47.

Batiste, W. C., Berlowitz, A., Olson, W. H., DeTar, J. E. & Joos, J. L. (1973). Codling moth: estimating time of first egg hatch in the field – a supplement to sex attractant traps in integrated control. *Environmental entomology*, **2(3)**, 387–91.

Betts, H. (1976). Forecasting infestations of tropical migrant pests: the Desert Locust and the African Armyworm. *Symp. Roy. Entomol. Soc. Lond.*, 7, 113–34.

Bowden, J. & Jones, M. C. (1979). Monitoring wheat bulb fly, *Delia coarctata* (Fallen) (Diptera: Anthomyiidae), with light-traps. *Bull. ent. Res.*, **69**, 129–39.

Croft, B. A., Howes, J. L. & Welch, S. M. (1976). A computer-based extension pest management delivery system. *Environmental entomology*, **5(1)**, 20–34.

Edwards, C. A. & Heath, G. W. (1964). *The Principles of Agricultural Entomology*. Chapman and Hall, London. 418 pp.

Fulton, W. C. & Haynes, D. L. (1977). The use of regression equations to increase the usefulness of historical temperature data in on-line pest management. *Environmental entomology*, **8(3)**, 393–9.

Gould, H. J. & Graham, C. W. (1977). The incidence of *Aphis fabae* Scop. on spring-sown field beans in south east England and the efficiency of control measures. *Plant Pathology*, **26**, 189–94.

Haynes, D. L., Bradenburg, R. K. & Fisher, P. D. (1973). Environmental monitoring network for pest management systems. *Environmental entomology*, **2(5)**, 889–99.

Hull, R. (1968). The spray warning scheme for control of sugar pest yellows in England: summary of results between 1959 and 1966. *Plant Pathology*, **17**, 1–10.

Joyce, R. J. V. (1953). Effect of the cotton plant in the Sudan Gezira on certain leaf-feading insect pests. *Nature*, **182** (4647), 1463–4.

Joyce, R. J. V. (1959). The yield response of cotton plant in Sudan Gezira to DDT spray. *Bull. ent. Res.*, **50**, 567–94.

Lea, A. (1958). Recent outbreaks of the brown locust, *Locusta pardalina* (Wlk.), with special reference to the influence of rainfall. *J. ent. Soc. S.* Africa, **21**, 162–213.

Odiyo, P. O. (1977). A forecasting system for the African Armyworm. *Proc. 3rd Scientific Meeting Int. Centr. Inst. Physiol. Ecol.*, **1976**, 60–66.

Smith, R. C. (1954). An analysis of 100 years of grasshopper population in Kansas (1854 to 1954). *Trans. Kans. Acad. Sci.*, **57**, 397–433.

Symmons, P. (1959). The effect of climate and weather on the numbers of the red locust, *Nomadacris septemfasciata* (Serv.), in the Rukwa Valley outbreak area. *Bull. ent. Res.*, **50**, 501–21.

Taylor, L. R. (1973). Monitor surveying for migrant insect pests. *Outlook on Agriculture*, **7**, 109–16.

Torii, T. (1971). The development of quantitative occurrence prediction of infestation by the rice stem-borer, *Chilo suppressalis* Walker, in Japan. *Entomophaga*, **16(2)**, 193–207.

Watson, M. A., Heathcote, G. D., Lauckner, H. B. & Sowray, P. A., (1975). The use of weather data and counts of aphids in the field to predict the incidence of yellowing viruses of sugar-beet crop in England in relation to the use of insecticides. *Ann. appl. Biol.*, **81**, 181–98.

Way, M. J. & Cammell, M. E. (1973). The problem of pest and disease forecasting – possibilities and limitations exemplified by work on the bean aphid, *Aphis fabae*. *Proc. 7th Brit. Insect. Fung. Conf.*, 933–54.

Way, M. J., Cammell, M. E., Alford, D. V., Gould, H. J., Graham, C. W., Lane, A., Light, W. I. St. G., Rayner, J. M., Heathcote, G. D., Fletcher, K. B. & Seal, K. (1977). Use of forecasting in chemical control of black bean aphid, *Aphis fabae* Scot., on spring-sown field beans, *Vicia fabae* L. *Plant Pathology*, **26**, 1–7.

Yasuo, S. (1971). Forecast work of disease and insect outbreak in Japan. *Japanese pesticide information*, **6**, 5–10.

PRACTICE OF PEST CONTROL

5 Physical Control of Pests

Physical control means the physical elimination of the pest or physical alteration of the environment to make it inimical or inaccessible to the pest. Such methods have been comparatively unimportant in modern agriculture, largely due to high labour costs. However, because of the side effects of insecticides, physical methods may serve as suitable control alternatives, especially when integrated into pest management programmes. Their use may be particularly attractive in developing countries where labour costs are comparatively low.

5.1 Methods of physical control

Physical control may be divided into two categories:
(1) Physical methods
(2) Environmental manipulation

5.1.1 Physical methods

Physical removal Hand picking of pests is hardly practical on crops grown on a large scale. But it is certainly useful in small farms, backyard gardens, etc. Removal of foci of infestation in store houses is known to greatly reduce and sometimes totally eliminate the insect population from the stored commodities. Similarly, primary hosts for field pests may be physically removed by various devices. In the control of citrus pests (Ebeling, 1959), physical measures such as jarring insects from trees, hand picking, destruction of egg masses, and removal by a strong stream of water still play an important role in the control of some species. With a thorough knowledge of the ecology of the pest it may be possible to develop novel physical control measures.

Use of drags This involves the use of brushes, chains or tarred paper which are dragged over the crops so that the pests are either crushed or stuck to the paper. These devices were commonly used before the advent of modern insecticides. Rolston & McCoy (1966) state that 'mechanical devices ranging from brush drags for knocking pea aphids from alfalfa to dozers for collecting grasshoppers are antiquated, but the fly swatter still persists'. Obviously their use needs to be reconsidered with a view to economizing the use of insecticides.

Barriers and adhesives These devices prevent the migration of pests and have considerable value in fruit growing. The trees are banded with a suitable barrier such as creosote, tar, lime, etc. Water is frequently used to exclude insects such as ants. Adhesives of various types, e.g. products based on hydrogenated castor oil, natural gum resin or vegetable wax, are known, but currently their principal use is for population surveys (see for example Prokopy, 1968).

Barriers such as sticky banding material, cotton or glass wool are frequently used to exclude insects from citrus crops (Ebeling, 1959). Cultivated units of this crop may also be surrounded by deep furrows or trenches in which crawling insects may be trapped and destroyed. Metal barriers have also been extensively used for this purpose.

A variety of physical barriers have been used to exclude insects from food stuffs. These may include the use of plastics and tins. Timber sheathing is used to prevent termite attacks. If properly installed, metal barriers around buildings effectively deter the subterranean termites (Bottrell, 1979). Underwater storage has also been considered with a view to exclude pests. Physical exclusion of pests under crop conditions has seldom received much attention though covering the crop with fine plastic mesh has been attempted to exclude insect pests. In the temperate regions, glass-houses and small plots have been successfully protected from insect pests by erecting barriers with mesh screens and aluminium foil casings.

Traps Non-toxic field traps were used for several decades in controlling economically important pests. Old fashioned traps include fly papers, sticky bands on fruit trees, beer-baited jam jars for cockroaches. The attractiveness of modern traps has been greatly improved by physical and/or chemical agents. For example, field traps with chemical attractants, attractant lights (ultraviolet radiation), release of carbon dioxide or a combination of these are now well documented (Nelson, 1967, Mazokhin-Porschnyakov, 1969). Eldumiati & Levengood (1972), working on the attractive response of moths to 357 nm radiation together with the resultant reduction of vigour and mating potential, suggest, that 'far-infrared' radiation can be used in light traps for reducing the population of economically important insects. The real value of traps, however, seems to lie in sampling and assessment of population of several pest species.

Instances of trapping as effective control agents on their own probably do not exist. But infestations of *Chilo tumidicostalis* have been reduced by light-trapping combined with collecting moths and egg masses by hand (Gupta & Avasthy, 1960). Practical application of insect attraction in the use of the light traps and areas of further research in this field are discussed by Harstock *et al.* (1966).

5.1.2. Environmental manipulation

The use of ecological factors against insects has been summarized by Banks (1976) who has reviewed post-1969 developments in this field and provided a bibliography of recent reviews. The use of these methods, as noted by Banks, is not as yet widespread against field pests and in stored commodities physical methods are of growing importance. In view of the problems posed by the increasing use of insecticides there is the need to re-examine every possibility in the light of modern technology (Busvine, 1968). The following control measures using ecological factors and their tolerance thresholds deserve more attention.

Dehydration Dust desiccants kill insects by destroying the water-proofing properties of their cuticles. This permits a lethal rate of water loss from the insect body. Substances used for this purpose include highly porous and finely-divided silica gel. Ebeling (1971) has reviewed their potential in pest control. Success in the control of cockroaches has been achieved by scattering diatomaceous earth (Boraiko, 1981) – a powder composed of sharp-edge shells of erstwhile freshwater microscopic plants – where roaches are most likely to pick up abrasive dust. Powdered boric acid has the same effect. While not as potent as synthetic insecticides, these cheap dusts are said to be longer lasting (Boraiko, 1981) and less repellent to the cockroach, an insect that learns to shun some chemical poisons before picking up a lethal dose.

Asphyxiation Control of insects by depriving them of oxygen has recently begun to receive fresh attention. It has been found that if insects and wheat grain can be sealed in an airtight container, by the time the insects have eaten about 150 g of wheat per 1000 kg they have so altered the atmosphere that life is no longer possible. This loss is less than one five thousandth of the weight of grain that has been destroyed before life can no longer continue. It is clear that if this were possible to achieve under conditions of field storage, the grain would always come out of storage without live insects, even when initially infested (see Cline & Highland (1978) for references on the subject). Airtight (hermetic) grain storage has been shown to be practical on a large scale under certain conditions (P.I.L., 1964).

Sound Many insects produce sounds, some have communication systems based on sound and quite a few, if not all, insect species respond to artificial sounds and vibrations. The practical use of sound has long been regarded as a possibility for insect control but so far its use as a practical control weapon has been largely unsuccessful. Its possible uses range from broadcast of alarm or alarming noises to disruption of sonic mate calling or use of sufficient sound volume to damage the insect (Frings & Frings, 1965, 1971). However, the most promising, from the control point of view, are the communication signals either recorded from the insects themselves or produced artificially. Three possibilities for using these signals are listed by Osmun (1972): (*i*) to attract insects in a trap, (*ii*) to repel them from specific areas, and (*iii*) to jam their natural communication systems. As noted by this author 'practical control for insects by sounds, even pilot-tests of practicality, remain for the future'.

Relative humidity The use of relative humidity in controlling insect populations in stored products is widely practised. The death of insects under a carbon dioxide-rich atmosphere is largely due to excessive water loss from insect bodies caused by prolonged opening of their spiracles. Howe (1965) notes that many storage pests cannot reproduce at ambient relative humidities of 50%. However, it is stated by Banks (1976) that, 'the possibility of producing an atmosphere dry enough to be insecticidal but which is only briefly out of equilibrium with the commodity merits further research'.

The manipulation of humidity in limiting insect populations in field crops has received little attention. McCoy (1962) notes that humidity imposes the main influence on adult activity of the fruit-fly, *Drosophila melanogaster*, where high levels of this factor favour oviposition in tomatoes. The potential of damaging populations of this fly can be reduced by maintaining tomato stands with sparse cover and a minimum of low-growing branches or other vegetation (Osmun, 1972). Much research with other insect species, in order to manipulate their habitats in such a way as to render humidity unfavourable for the development of an economically damaging population, still needs to be carried out.

Temperature High temperatures (45°C), normally used in artificially drying grains for storage, for feed pellets and food-processing facilities, are lethal to insects. Efficient, portable, grain-drying equipment has been used for this purpose as well as the use of special elevator heating pipes in store houses. Among other examples, Osmun (1972) mentions the control of insects in baled fibres and logs being seasoned for lumber; insects and nematodes in soil and insects in special commodities and under special conditions such as in bulbs, grapes, oranges, and house plants. Flaming (the use of liquid fuel flame) has in recent years been used in the eastern United States to reduce alfalfa weevils (Pass *et al.*, 1967). Propane flamers destroy any weevil eggs and adults in alfalfa stubble. However, there are obvious limitations to the use of temperature in controlling insects as many plant materials themselves are unable to withstand high temperatures.

Flaming also destroys natural enemies and other beneficial organisms (Flint & van den Bosch, 1977) and requires petroleum fuel. On the other hand, lowering of temperature retards the development of insects. The reduction of commodity temperature is widely used in the protection of grain and other stored products (see e.g. Navarro *et al*., 1969; and Bottrell, 1979). In welldesigned stores, natural cooling can be effected by ventilation. Cold temperature is also used as a quarantine measure against some pests of export fruit. In the United States, fruits imported from countries infested with the Mediterranean fruit-fly, must be refrigerated for 12–20 days depending on the fly species involved. This treatment kills the immature stages of the fruit-fly.

Electro-magnetic energy These methods have been studied largely in connection with stored commodities and may be considered under the following sections. For a recent review of the subject reference should be made to the article by Nelson (1973).

(1) *Longer wave-length energy* ($10^3 - 10^6$ nm)
This form of energy can only be used in the form of very short radio waves or infrared radiation which can be transformed into heat. But the costs of this method are considerably higher than conventional control methods (Busvine, 1968).

(2) *Medium wave-length energy* ($10 - 10^2$ nm)
Light and ultraviolet radiations are frequently used to attract insects to traps as mentioned earlier (see Banks, 1976 for useful references and a summary of the subject). Recently it has been shown that a few, very bright, flashes of ultraviolet light may affect the endocrine systems of insects. For example such exposures can prevent insects entering diapause. Further, radiation may have deleterious effects on the growth of some insects following daily exposure of a few minutes to red and infrared radiation (Busvine, 1968). The full potential of light induced phenomena in pest control has yet to be realised and requires much further basic research. As noted by Osmun (1972), 'no practical control measures involving light regulation and its subsequent effect in insect population control are at present being practiced'.

(3) *Ionizing radiation* (1 nm and less)
Atomic radiations readily penetrate animal tissues though

insects are far more tolerant of them than vertebrates. Tolerance levels for several insects have been worked out (see Watters, 1968 for a review) and with a suitably guarded source, such as radio-active cobalt or an X-ray machine, grain can be disinfected at the rate of several tonnes per hour. However, an irradiated commodity is liable to reinfestation and such a method is therefore unsuitable for unpackaged goods. The use of atomic radiation to control insects by 'male sterility' is discussed elsewhere (Chapter 9).

5.2 Future of physical methods in pest control

Since physical methods are free from the drawbacks experienced with the use of pesticides (see later), there is the need to re-examine every technique for the control of pests in the light of modern technology. Even largely unsuccessful methods e.g. the practical use of electric discharge as a control method (Anonymous, 1969), physical shock (Bailey, 1969), etc. need to be re-explored in detail. Widely studied methods, such as environmental gas control (e.g. Bailey & Banks, 1975) need to be studied in greater detail. Methods of physical control without doubt provide very useful tools in integrated pest management strategies (see Chapter 13), and with an increasing awareness of the biology of the pest it may be possible to develop unique physical methods of pest control. The utilization of physical approaches as 'restraining agents' in pest control is viewed with moderate optimism by Osmun (1972). In Africa some physical practices such as the removal of foci of infestations in stores and the use of refrigeration to control pests in households are currently in use. Hand picking of egg masses of stem-borers is also practised but little else.

5.3 Literature cited

Anonymous (1969). Insect-o-cutor (use against flies in poultry houses). *PANS*, **15**, 280.

Bailey, S. W. (1969). The effects of physical stress in the grain weevil, *Sitophilus granarius. J. stored Proc. Res.*, **5**, 311–25.

Bailey, S. W. & Banks, H. J. (1975). The use of controlled atmospheres for the storage of grain. *Proc. 1st Int. Working Conf. Stored Prod. Ent., Savannah, USA*, 362–74.

Banks, H. J. (1976). Physical control of insects – recent developments. *J. Aust. ent. Soc.*, **15(1)**, 89–100.

Boraiko, A. A. (1981). The indomitable cockroach. *National Geographic Magazine*, **159(1)**, 130–42.

Bottrell, D. R. (1979). *Integrated Pest Management.* Council on environmental quality. U.S. Government Printing Office, Washington, D.C. 120 pp.

Busvine, J. R. (1968). The future of pest control. Part II. Alternatives to insecticides. *PANS*, **14(3)**, 318–28.

Cline, L. D. & Highland, H. A. (1978). Survival of four species of stored product insects in airtight laminated food pouches. *J. Econ. Ent.*, **71(1)**, 66–8.

Ebeling, W. (1959). *Subtropical Fruit Pests*. University of California, Division of Agricultural Science, Berkeley, 436 pp.

Ebeling, W. (1971). Sorptive dusts for pest control. *Ann. Rev. Ent.*, **16**, 123–58.

Eldumiati, I. I. & Levengood, W. C. (1972). Summary of attractive responses in Lepidoptera to electromagnetic radiation and other stimuli. *J. Econ. Ent.*, **65(1)**, 291–2.

Flint, M. L. & van den Bosch, R. (1977). A source book on integrated pest management. *Univ. Calif. Int. Center for Integrated Biol. Control* 392 pp.

Frings, H. & Frings, M. (1965). Sound against insects. *New Scientist*, **26** (446), 634–7.

Frings H. & Frings M. (1971). Sound production and reception by stored products insect pests – a review of general knowledge. *J. Stored Prod. Res.*, **7**, 153–62.

Gupta, E. D. & Avasthy, P. N. (1960). Biology and control of the stem borer *Chilo tumidicostalis. Proc. Int. Soc. Sug. Cane Technol.*, **10**, 886–901.

Harstock, J. G., Deay, H. D. & Barrett, J. R., Jr. (1966). Practical application of insect attraction in the use of light traps. *Bull. Ent. Soc. Amer.*, **12**, 375–7.

Howe, R. W. (1965). A summary of estimates of optimal and minimal conditions for population increase of some stored product insects. *J. Stored Prod. Res.*, **1**, 177–84.

Mazokhin-Porschnyakov, G. A. (1969). *Insect Vision*. Plenum Press, New York. 306 pp.

McCoy, G. E. (1962). Population ecology of the common species of *Drosophila* in Indiana. *J. Econ. Ent.*, **55**, 978–85.

Navarro, S., Donahays, H. & Calderson, M. (1969). Observations on prolonged grain storage with forced aeration in Israel. *J. Stored Prod.* Res., **5**, 73–81.

Nelson, S. O. (1967). Electromagnetic energy. pp. 89–145. In *Pest Control: Biological, Physical and Selected Chemical Methods.* Ed: Kilgore, W. & Doutt, R. L. Academic Press, London. 206 pp.

Nelson, S. O. (1973). Insect control studies with microwaves and other radiofrequency energy. *Bull. Ent. Soc. Amer.*, **19**, 157–63.

Osmun, J. V. (1972). Physical methods of pest control. *J. Environ. Qual.*, **1(1)**, 40–5.

Pass, B. C., Smith, E. M., Templeton, W. C. & Cook, D. (1967). Control of the alfalfa weevil in Kentucky using LP-gas flame. *Proc. 4th Symp. thermal Agric.*, 38–41.

P.I.L. (1964). *Report of the Pest Infestation Laboratory for 1964*. p. 19. Agricultural Research Council, U.K.

Prokopy, R. J. (1968). Sticky spheres for estimating apple maggot adult abundance. *J. Econ. Ent.*, **61**, 1082–5.

Roiston, L. H. & McCoy, C. E. (1966). *Introduction to Applied Entomology*. The Ronald Press Company, New York. 208 pp.

Watters, F. L. (1968). An appraisal of gamma irradiation for insect control in cereal foods. *Manitoba Entomol.*, **2**, 37–45.

6 Cultural Practices

Methods of pest control by manipulating crops and land are among the oldest traditional practices. They are concerned with making the environment unfavourable for the pest and thereby either averting the damage or at least limiting its severity. They can greatly influence the level of a pest's field population by killing the pest or adversely affecting its fecundity. Or, alternatively, they may provide a more favourable environment for the natural enemies of the pest. In all these instances, a thorough knowledge of the life history and habits of the insects and its plant host may be required.

The role of cultural practices as an economic form of pest management seems to have been little appreciated. But the use of cultural practices inimical to insect development has figured prominently in the control of many serious pests. While it is true that these practices are generally labour intensive and the high cost of labour in the developed world has led to their abandonment, there is no reason why they should not be pursued vigorously in the developing countries where labour is surplus. Though rarely spectacular in their effect, the practices are nevertheless dependable and even a partial control may reduce dependence on insecticides, thereby assisting in our goal of establishing modern ecologically-sound pest-management systems.

Cultural practices are readily available to the rural farmer and in most cases do not entail any extra investment in equipment to carry out insect control. It is easy to imagine, for instance, that one of the advantages of shifting agriculture and mixed cropping in Africa is the avoidance of high losses from pests, disease and weeds. Slight improvement of cultural practices may, in conjunction with other control measures, greatly increase the effectiveness of the general pest control programme.

6.1 Economics of cultural control

Very few cost/benefit analyses of the use of cultural practices anywhere in the world are available. In the United States the Environmental Pollution Panel of the President's Science Advisory Committee (1965) has estimated that the use of the cultural practice of 'delay seeding until the fall brood has disappeared', in controlling Hessian fly on wheat, brings a return of \$55 for each dollar invested in research whereas the return for chemical treatment is less than \$2 for every \$1 investment. The details at national level are as follows:

Losses if no control measures are adopted

22% loss, i.e. 83 600 000 bushels of wheat per year or \$167 200 000 (average value \$2.00 per bushel) + \$24 000 000 loss per year in grazing value (4 000 000 acres at \$6.00 per acre).

Total losses in absence of control measure = \$191 200 000.

Cost/benefit of control by delayed seeding

Loss = 20 000 000 acres (yielding 380 000 000 bushels of wheat) × 7% loss × bushel yield × \$2 per bushel or \$53 2000 000 + \$24 000 000 loss in grazing value.

Total loss = \$77 200 000

Net gain = \$191 200 000 – \$77 200 000 = \$114 000 000 a year.

Costs of developing the delayed seeding method

10 man-years per year (5 entomologists + agronomists) for 20 years (1915–1935) at a cost of \$10 000 per year.

Total cost = \$2 000 000.

Cost of maintaining an information service to farmers, warning them of fly-free date each year = \$20 000. Therefore, there is a potential saving of \$114 200 000 per year for a total research cost of \$2 000 000 and a maintenance cost of \$20 000 per year, i.e. \$55 per year for each dollar invested in research and extension.

Cost/benefit of control by chemicals

Cost of treatment for 20 000 000 acres at \$4 per acre = \$80 000 000 + 4% loss by spring brood (total expected yield = 380 000 000 bushels of wheat at \$2.00 per bushel) = \$30 400 000 + loss of grazing value = \$24 000 000

Total cost and loss = \$134 400 000

Saving as a result of chemical treatment = \$191 200 000 – \$134 400 000 = \$56 800 000

Therefore return is less than $2 for each dollar invested.

From the foregoing example and from the account given below the role of cultural practices in effecting a substantial saving to world agriculture appears immense.

6.2 Some common cultural practices

6.2.1 Destruction of residues, alternative hosts and volunteer plants

Practices such as the destruction of the stubs of maize and sugar-cane lessen carry-over of diseases or pests from one season to another. Burning, flooding or ploughing stubble after the crop has been harvested reduces the borer attack in the next season.

The effects of stalk and boll destruction and tillage as well as the influence of early crop termination on the number of larvae of cotton pink bollworm (*Pectinophora gossypiella*) entering diapause have been experimentally investigated in the United States. Ploughing tests and winter burial studies by Adkisson *et al.* (1960) have indicated that 76–83% of the pink bollworm larvae in infested cotton debris are killed by burying the material during winter. These authors also found that stalk-shredding machines could kill up to 88% of the larvae in the field after harvest.

In Ghana and Nigeria, field observations have shown that if maize stalks remain in the field after harvesting, borer incidence increases considerably when maize is grown in both first and second seasons on the same land in consecutive years. The stalks contain larvae and pupae of the stem-borers and adults emerge from the stalks and re-infest any young maize plants. The recommended practice of burning stalks completely after the grain has been harvested with a view to killing all the diapausing larvae, is seldom followed by the farmers who use stalks for a variety of purposes. The recent suggestion of Adesiyun & Ajayi (1980) of partial burning of stalks which results in killing 95% of the larvae still needs to be adopted by the West African farmers.

In Ghana, *Eldana saccharina* is a major pest of sugar-cane stands. Studies have shown that the larvae of *E. saccharina* remain in the stubble after harvest and re-infest new shoots.

Thus the stubble serves as a reservoir for re-infestation by stem-borers (Kumar & Sampson, 1983). The stubble larvae destroy the shoots in the ratooned field thereby reducing the tillers of the ratoons. The idea of discontinuing the practice of ratooning with a view to remove the stubble which serves as a reservoir for re-infestation still remains to be adopted by the sugar-cane estates.

In the Sudan the destruction of crop residues and weeds has been considered essential in preventing infestation by pink bollworms. It is also regarded as an important part of the control measures against blackworm disease and leaf curl disease (Ripper & George, 1965). In Sudan, all cotton plants, whether volunteer standover or abandoned cotton must be destroyed after harvest. This practice of phytosanitation is enforced by legislation. In Nigeria, to control pink bollworm, it is illegal to grow cotton during the cotton 'closed season' which is enforced annually from 1st March to 1st July in the north and from 15th March to 15th June in the south. During this period, all cotton plants are cut off at ground level or uprooted and burnt as soon as harvesting has been completed. All growth from old stumps and waste is also destroyed. Similar legislation has been enacted in Kenya.

In Botswana the destruction by burning of *all* cereal residues, immediately after harvest, is strongly recommended in order to control the stalk-borers such as *Chilo* and *Busseola* (Ingram *et al*., (1973). These workers also advocate the collection and burning of all discarded blind heads and seeds of sorghum to cut down the incidence of sorghum midge, *Contarinia*. Similarly for the control of the *Bagrada* bug, complete destruction of all residues of the crucifers has been recommended. This also applies to an early crop that has failed due to a mid-season drought – it should be ploughed in before re-sowing (Ingram *et al*., 1973).

Decayed and dropped fruits usually offer refuge to the developing stages of their pests. For example, in both East and West Africa any ripe or dried coffee berries left on the ground or lost in mulch during harvesting act as breeding sites for the berry-borer, *Hypothenemus hampei*.

In West African cocoa farms removal of all old pods, after harvesting, whether diseased or not, has been strongly recommended. This method eliminates pods which can act as a source of inoculum for the spread of the black-pod disease. It also helps to break the life cycle of insect pests such as moths and certain bugs (*Bathycoelia* spp.). Also old pods, together with dead twigs, are used as nest sites by *Pheidole* and *Crematogaster* ants. The latter have been considered to be the most important vectors of black-pod disease. Infested pods left on the ground after harvesting or left hanging on the tree are used as building materials by the ants which therefore effectively transfer infected pod material to healthy pods. Rotting vegetables left in the field are frequently infested with caterpillars. Their destruction undoubtedly lessens the carry-over of the pests.

In the United States the spittle bug (*Prosapia bicincta*) feeds on Bermuda grass, a high quality forage, and Beck (1963) established complete control by burning, in early April, all refuse and dead grass remains from the previous year.

6.2.2 *Dates of sowing and harvesting chosen to avoid pest attack*

Control of some insect pests is achieved by following the principle of growing the crop when the pest is not present or of planting so that the most susceptible stage of a crop development coincides with the time when the pest is least abundant.

Groundnut (*Arachis hypogaea*) in Africa south of Sahara suffers considerable losses from rosette virus disease transmitted by *Aphis craccivora*. However, it has been found that groundnuts sown early in the wet season are old and less attractive at the time of aphid invasion, and suffer less from attack than those planted later. (See Farrell, 1976 for references.) In both Nigeria (Booker, 1963) and Tanzania (Evans, 1951), early planting of this crop has also been shown to produce higher yield than late planting.

Endrody-Younga (1968) carried out an extensive study of maize infestation by the stem-borer, *Sesamia botanephaga*, in Central Ashanti (Ghana). He concluded that maize planted at the start of the first rainy season (February through April), without fertilizer use, insecticide application or irrigation, gave yields estimated higher than the average given by FAO (above 1000 kg ha^{-1}) for the country as a whole. Thus the cultural practice of growing maize at the beginning of the first rainy season to avoid borer attack does not necessitate any control measures against the borer pests. However, to obtain a meaningful crop during the second season, required chemical control of the borers, the use of fertilizers and irrigation of land.

In the Sudan savanna zone of Northern Nigeria, sorghum and millets are sown by farmers as early as possible in the wet season. Yields are greater when an early onset of season is accompanied by early sowing. Further, cultivars used by farmers in this region have been selected over long periods 'to start heading so as to set seed at the end of the normal rainy season – with the result that grains are not attacked by moulds and insects as they would be if heading was earlier' (Kassam *et al.*, 1976).

In Sudan, Ripper & George (1965) state that to obtain greater yields from the cotton crop, planting must be early and this also, incidentally, decreases the incidence of pink bollworm and the spiny bollworm. At Morogoro, Tanzania, Enyi (1974) observed that early planting of cowpea gave a much higher yield than late planting, without any attempt at insect control. Planted in January, the unsprayed crop produced almost as high a yield as the sprayed crop since the young seedlings established themselves before *Ootheca benningseni* attack and matured before *Acanthomia* sp. became abundant.

It is, however, not always true that early planting reduces pest abundance, for this depends on the life history of the insect in question, the crop and its location. It is often possible to avoid attack by planting a crop well *after* the time of egg laying by the pest. The hessian fly (*Mayetiola destructor*) normally emerges in the autumn and lives for only 3–4 days. If the winter wheat is planted when most of this generation is past, the plants will have few eggs laid on them (Anonymous, 1969). As mentioned earlier, entomologists in the hessian fly-infested areas of the United States have established dates for sowing winter wheat that will allow the plants to make satisfactory autumn growth but should be late enough to avoid heavy hessian fly infestation.

In China, farmers have from time immemorial used careful selection of sowing dates to prevent infestation by rice stem-borers.

The utilization of host-free periods has been of great

importance in controlling subtropical fruit pests (Ebeling, 1959). A host-free period has resulted in the reduction to negligible proportions of the Mexican (black) fruit-fly (*Anastrepha ludens*) population in the Rio Grande Valley of Texas. The inauguration of a host-free period has also been considered an important element of success in controlling the Mediterranean fruit-fly (*Ceratitis capitata*) during the Florida eradication campaign (1929–30).

As with the time of planting, so the time of harvesting may have a pronounced affect on the insect population of a field. Early cutting of the first and second crops of alfalfa has been stated (Anonymous, 1969) as a practical control method for the alfalfa weevil, *Hypera postica*. Similarly cotton growers in the United States endeavour to mature a crop before a population build-up of the boll weevil can do extensive damage (Anonymous, 1969). Much, however, remains to be learnt in Africa about the use of harvesting time in containing damage by insect pests.

6.2.3 Good husbandry

The value of good husbandry in crop protection cannot be over-emphasized, especially for tree crops. For example, neglect of pruning in coffee farms is known to encourage the incidence of the berry borer (*Hypothenemus hampei*) because natural shade is increased and berries are left unpicked on the tall trees which result. Since the pest will breed only in hard beans, regular picking of ripe berries at least every two weeks will remove many of the breeding insects and much of the available food. These methods of crop hygiene, together with reduction of shade, will usually keep berry-borer populations at a low level (McNutt, 1975).

In cocoa farms the effectiveness of removing old pods, epiphytes, climbers and mistletoes, of dead wood from both trees and the ground, brushing etc. is undeniable. Most farmers are neglectful in this respect. Mistletoes are responsible for thinning of the cocoa canopy and this invites destructive mirid invasions. Epiphytes, climbers and mistletoes provide nest sites for unwanted ant genera such as *Pheidole* and *Crematogaster. C. striatula* is positively associated with mealybugs which are vectors of swollen shoot disease. Removal of old pods, etc. is therefore of importance in limiting both black-pod and swollen shoot disease. Good husbandry is often cheap and effective and should not be neglected in favour of other methods of control.

In the Indian Tea Gardens, red spider mite (*Oligonychus coffeae*) can be an important pest and persists on a few old leaves as well as on the scale leaves, at the base of shoots, during cold weather. Removal of these leaves during the lean period greatly reduces the incidence of the mite on the crop (Das, 1960).

Poor weed control often drastically reduces crop yields as a result of competition for light, nutrients, moisture and space. The weeds also act as alternate hosts for a variety of pests and their destruction is known to prevent damage to future crops and assist in the control of certain insect pests (Barnes, 1959; Anonymous, 1969). In Northern Nigeria the vector of the groundnut rosette virus disease, *Aphis craccivora*, occurs on the weed *Euphorbia hirta* throughout the dry season. Up to 40% of this common weed has been found to be infested at the end of the dry season. Destruction of the weed has been recommended in order to reduce the aphid population transmitting the disease in the following growing seasons. In Botswana the destruction of weeds has been found to markedly reduce the amount of damage to crops such as cotton, cowpea, sorghum and maize caused by the leaf-eating weevils, *Protostrophus* spp. (Ingram *et al.*, 1973).

Experiments in Nigeria have shown (IITA, 1973) that weeds in grain legume fields are positively correlated with insect damage to cowpea and soybean. The unweeded plots of specified cowpea and soybean varieties showed higher percentage of seed damaged by hemipterous bugs and the moth *Laspeyresia* (Table 1).

Table 1 Effect of weeds on legume pod and seed pests in the field (IITA, 1973, modified).

Legume	**Weeds**	**% of seeds damaged by**		
		Maruca	**Hemiptera**	***Laspeyresia***
Cowpea var. Prima	Absent	1.6	27	18
	Present	2.1	58	26
Soybean var. Kent	Absent	0	46	6.5
	Present	0	59	8.3

In sugar-cane fields the elimination of weeds may help to reduce moth borer infestations if grasses are important alternative hosts. For example, in northern Queensland, the large moth borer, *Bathytricha truncata*, causes serious losses to sugar-cane fields that are allowed to become grassy due to neglect during the growing season.

Most of Africa's better farmers now have advanced beyond hand broadcasting of their seeds and sow in rows. Row cropping also makes weeding and cultivation easier. However, the importance of weeds and undergrowth must be worked out for each crop-pest situation. Thus Way (1953) found that the coreid bug, *Pseudotheraptus wayii*, the cause of premature coconut fall in Zanzibar and the coastal areas of Tanzania, is controlled to some extent by the ant, *Oecophylla longinoda*, which thrives in unweeded plantations where undergrowth provides a favourable habitat.

6.2.4 Rotation of crops to avoid build up of pests

This practice is most effective against insects having a restricted host range and those possessing limited migratory capabilities. Rotation cropping has been effective in minimizing the attack of nematodes in tomatoes and cowpeas. Rotation of groundnuts with crops non-susceptible to nematodes is considered the most effective of all methods in reducing the numbers of pest nematode in the soil. On the other hand, control of root nematodes in rice by crop rotation is unlikely to be effective owing to the wide host range of root knot nematodes.

The most common rotations involve grasses, legumes and root crops. Insects that feed on grasses seldom patronize legumes and roots. For example, some leguminous crops are unfavourable to the development of white grubs that feed on the roots of many crops. In the U.S.A., the proper use of legumes in rotation with grass crops has been found to greatly reduce white grub injuries (Anonymous, 1969). While it is difficult to lay down hard and fast rules, it is best to avoid a sequence of closely-related crops. In planning a suitable crop rotation exercise, a knowledge of the life cycles of the pests and their hosts within a particular area is required. For an account of rotation of vegetable crops in the tropics reference should be made to the article by Adjei-Twum (1971).

6.2.5 Cropping systems

Okigbo (1978) states that the most widespread cropping system in Africa consists of mixed intercropping in which several species of crop plants (both annuals and perennials) develop in compound farms to form a complex and fairly stable agro-ecosystem. He concludes that the traditional farmer practises intercropping because it gives higher total yields and greater returns than the same crops grown in pure culture. The advantages of mixed cropping under low level inputs have also been discussed by Norman (1974) and Finlay (1975). It is interesting to note that in West Africa the insect-susceptible grain legumes, such as cowpea, are generally mixed with other crops (cereals, cotton, pepper, etc.) while the more resistant grain legumes such as soybean, bambarra groundnut (*Voandzeia subterranea*) and groundnut are often planted as a single crop. It is true that the yield of cowpea grown as a sole crop and sprayed with insecticide is much higher than that of intercropped but unsprayed cowpea. But in the absence of effective pest-control programmes, a traditional farmer is assured of at least some yield of the crop he is interested in when he practises intercropping.

IRRI (1973) reported that intercropping of groundnuts and maize markedly decreased corn borer (*Ostrinia furnacalis*) damage in the maize. This has been attributed to increased activities of spider (*Lycosa* spp.) predation in the presence of groundnut.

Guillemin (1952), in the Central African Republic, found that groundnut intersown with maize suffered less rosette infection than did groundnut in monoculture. In two field trials in Malawi, Farrell (1976) found that the spread of rosette virus disease in groundnut intersown with field bean was less than the spread in groundnut monocultures. The coreid bug, *Anoplecnemis curvipes* which damages cowpea pods has been found to exhibit greater preference for egg laying on maize than on cowpea (IITA, 1975). Singh *et al.* (1978) reported observations to the fact that this bug does not lay its eggs on cowpea plants but will lay on maize planted with cowpea. Further, the nymphs of *A. curvipes* are known to cause no visible damage to maize by their feeding. Singh *et al.* (1978) reported that the blister beetle, *Mylabris farquharsoni*, can totally destroy one cowpea flower in about a minute. If maize tassels are present in the mixed crop at the time of cowpea flowering all the beetles are attracted to the tassels. Cowpea flowers suffer heavy damage in the absence of maize plants. It is difficult to control this pest with insecticide sprays and the beetles feed on flowers that last only a day.

A well-known example of an insect pest whose damage is reduced by intercropping is the coreid bug, *Pseudotheraptus wayii*, mentioned earlier. It has been found that the intercropping of cashew or citrus with coconut also favours the predator in the same way that the undergrowth does in a neglected coconut plantation (Kayumbo, 1977).

Okigbo (1978) believes that as a result of minimized pest and disease losses and losses due to adverse environmental conditions, risk is lower in intercropping than sole cropping. Way (1976) also observes that African farmers, 'have evolved over centuries, methods of multiple and intercropping which, besides other advantages, possess certain empirically – developed elements of insect pest control'. Way (1976) believes that the other crops in a mixed cropping system may limit pest incidence in one or more ways, including the following:

(1) 'by acting as a barrier, or a hazard, or a camouflage;
(2) by acting as alternative hosts, diverting the pest away from the crop at risk;
(3) by benefiting natural enemies of the pest'.

However, much remains to be learnt about the range of interactions between pests and crops and the natural control mechanisms operating in the mixed or multiple cropping systems practised in Africa. Extensive studies of farmers' fields are especially required.

6.2.6 *Plant density*

Observations and experiments indicate that plant density may influence the incidence of certain pests in some crops. Booker (1965), in Northern Nigeria found that closely-spaced cowpea had significantly more beetles per acre than widely-spaced cowpea throughout the period tested. The ratio approached 2:1 in most weeks (Table 2), though there was a tendency towards the ends of the counts for this ratio to decrease. On the other hand, close spacing of groundnut merely reduces the percentage

Table 2 Populations of *Ootheca mutabilis* on cowpea grown at two spacings at Samaru, Nigeria (Booker, 1965).

Weeks after planting	Spacing (cm) 46 (18″)	91 (36″)
2	712	530
3	2164	1024
4	1819	1019
5	773	419
6	609	381
7	251	132
Mean	1067	548

of rosette-infected plants but not the number of diseased plants per unit area (Booker, 1963). This agrees with work in Tanzania (Evans, 1954) and Senegal (Tourte & Fauché, 1954). Results of A'Brook's (1968) investigations indicate that the control of groundnut rosette disease by early planting and close spacing arises from the fact that few vectors land in close-spaced groundnut crops. It appears that continuous groundnut cover when the vector migrates into the crop inhibits the landing response and thereby produces control of the rosette virus transmitted by aphid vectors such as *Aphis craccivora*.

In wheat Luginbill & McNeal (1958) found that the infestation and cutting by the wheat-stem sawfly, *Cephus cinctus*, decreased as row spacing became less. The sawflies apparently selected the larger, more succulent stems in the lighter seedling for oviposition. It would appear, therefore, that the role of spacing or plant density in insect control depends on crops, type of insects and other factors. In most crops in Africa this role has either not been investigated or not clearly understood.

6.2.7 Other practices

Other cultural practices available to the farmer are the use of tillage, improved storage and processing, strip-cropping, maintenance of plant vigour by proper irrigation, fertilizer use, use of 'trap-crops', picking of egg masses etc. Sometimes the depth of planting also affects the incidence of some pests: Akhade *et al.* (1970) report that by planting seed potatoes at 10 cm below soil level instead of 6 cm, infestation due to potato tuber moth (*Pthorimaea operculella*) was reduced from 18.1% to 9.1%.

Cultural practices are liable to variation in accordance with the particular requirements of the local conditions as well as the nature of the crop. Williams (1947), some thirty years ago, in his address on 'the field of research in preventive entomology' emphasized the need for greater effort in research on fundamental problems. Nowhere is there more need for work than on such fertile problems as the development of novel cultural practices in the developing world. The emphasis must be on continually altering the environment to suit the needs of man. Advantage should be taken of the 'weak-links' (Adkisson, 1977) in the seasonal cycle of the pests with a view to their suppression and at the same time conservation of natural enemies. For this purpose a thorough knowledge of the pest complex of the crop as well as the effect of crop phenology on the pest and its natural enemies is required (Adkisson, 1977).

It is interesting to note that the Entomological Society of America's special publication of August 1975 on *Integrated Pest Management* makes the following recommendation in respect of cultural control:

(1) 'Develop mechanisms for interdisciplinary evaluation of various cultural controls with a view to inclusion in integrated pest management systems so as to greatly reduce or eliminate the initiation of counter-productive practices'.
(2) 'Encourage extension services to reemphasize cultural controls as an effective, economical across-the-board tactics for integration into pest management systems'.
(3) 'Expand and intensify research on the methodology of producing pest-free seed and *seed-stocks* (apparently meaning planting material) and intensify education efforts to convince growers of the importance of using such seed'.

6.3 Literature cited

A'Brook, J. (1968). The effect of plant spacing on the numbers of aphids trapped over the groundnut crop. *Ann. appl. Biol.*, **61**, 289–94.

Adesiyum, A. A. & Ajayi, O. (1980). Control of the Sorghum stem borer, *Busseola fusca*, by partial burning of stalks. *Tropical Pest Management*, **26(2)**, 113–17.

Adjei-Twum, D. C. (1971). Rotation of vegetable crops in the tropics. *World Crops*, 10–15.

Adkisson, P. L. (1977). In *Proceedings of the UC/AID* – University of Alexandria, A.R.E. seminar/Workshop in Pesticide Management. Alexandria, Egypt. 175 pp.

Adkisson, P. L., Wilkes, L. H. & Cockran, B. J. (1960). Stalk shredding and ploughing as methods for controlling pink bollworm, *Pectinophora gossypiella. J. Econ. Ent.*, **55(3)**, 436–9.

Akhade, N. N., Tidke, P. M. & Patkar, M. B. (1970). Control of potato tuber moth, *Gnorimoschema operculella* Zell. in Deccan plateau through insecticides and depth of planting. *Indian Journal of Agricultural Science,* **40(12)**, 1071–6.

Anonymous (1969). Principles of plant and animal pest control. Vol. 3. *Insect–Pest Management and control – Cultural Control* pp. 208–42. National Academy of Sciences, Washington D.C. 508 pp.

Barnes, O. L. (1959). Effect of cultural practices on grasshopper populations in alfalfa and cotton. *J. Econ. Ent.*, **52**, 336–7.

Beck, E. W. (1963). Observation on the biology and cultural – insecticidal control of *Prosapia bicincta*, a spittle-bug, on Coastal Bermuda grass. *J. Econ. Ent.*, **56(6)**, 747–52.

Booker, R. H. (1963). The effect of sowing date and spacing on rosette disease of groundnut in Northern Nigeria with observations on the vector, *Aphis craccivora. Ann. Appl. Biol.*, **52**, 125–31.

Booker, R. H. (1965). A note on the effect of spacing of cowpea on the incidence of *Ootheca mutabilis* Sahlb. (Chrysomelidae). Miscellaneous Paper no. 10. Institute for Agricultural Research, Samaru, Nigeria. 2 pp.

Das, G. M. (1960). Occurrence of the red spider, *Oligonychus coffeae* (Nietner), on tea in North-East India in relation to pruning and defoliation. *Bull. ent. Res.*, **51**, 415–26.

Ebeling, W. (1959). *Subtropical Fruit Pests*. University of California, Division of Agricultural Science, Berkeley. 436 pp.

Endrody-Younga, S. (1968). The stem borer, *Sesamia botanephaga* Tams and Bowden (Lep., Noctuidae) and the maize crop in central Ashanti, Ghana. *Ghana Jnl. agric. Sci.*, **1**, 103–31.

Enyi, B. A. C. (1974). Effects of time of sowing and phosphamidon (Dimecron) on cowpea (*Vigna unguiculata*). *Expl. Agric.*, **10**, 87–95.

Evans, A. C. (1954). Groundnut rosette disease in Tanganyika. I. Field studies. *Ann. appl. Biol.*, **41**, 189–206.

Farrell, J. A. K. (1976). Effects of groundnuts sowing date and plant spacing on rosette virus disease in Malawi. *Bull. ent. Res.*, **66**, 159–71.

Finlay, R. C. (1975). Intercropping soybeans with cereals. In *Soybean Production, Protection, Utilization*. Ed. D. Whigham – INISOY series no. 6. University of Illinois, Urbana, U.S.A., pp. 77–85.

Guillemin, R. (1952). Études agronomiques sur l'arachide en A.E.F. Oubangui-Chari, Tchad. *Oléagineux*, **7**, 699–704.
IITA (International Institute of Tropical Agriculture) (1973). Report – Grain legume improvement program, Ibadan, Nigeria. **1972**, 31–2.
IITA (International Institute of Tropical Agriculture) (1975). Annual Report, 1974. Entomology, p. 101.
Ingram, W. R., Irving, N. S. & Roome, R. E. (1973). *A Handbook on the Control of Agricultural Pests in Botswana*. Government Printer, Gaborone. 129 pp.
IRRI (International Rice Research Institute) (1973). Annual Report, 1972. Intensification of cropping-weed control – insect management. Los Baños, Laguna, Phillipines. **1972**, 27–31.
Kassam, A. H., Dagg, M., Kowal, J. M. & Khadr, F. H. (1976). Improving food crop production in the Sudan Savanna zone of Northern Nigeria. *SPAN*, **8(6)**, 341–7.
Kayumbo, H. Y. (1977). Ecological background to pest control in mixed cropping systems. In *Proceedings of the Workshop on Cropping Systems in Africa. Morogoro, Tanzania, 1–6 December*, 1975, 51–3.
Kumar, R. & Sampson, M. (1983). A review of stem borer research in Ghana. *Insect Sci. Application* (in press).
Luginbill, P., Jr. & McNeal, F. H. (1958). Influence of seeding density and row spacings on the existence of spring wheats to the wheat stem sawfly. *J. Econ. Ent*., **51(6)**, 804–8.
McNutt, D. N. (1975). Pests of coffee in Uganda, their status and control. *PANS*, **21(1)**, 9–18.
Norman, D. W. (1974). Crop mixture under indigenous conditions in the northern part of Nigeria. *Samaru Research Bulletin,* **205**. Institute for Agricultural Research, Samaru, Nigeria. pp. 130–144.
Okigbo, B. N. (1978). Cropping systems and related research in Africa. *Association for the Advancement of Agricultural Sciences in Africa (AAASA). Occasional Publications Series – OT-1*. 81 pp.
President's Science Advisory Committee (1965). *Restoring the quality of our environment*. Report of the environmental pollution panel. The White House. U.S. Government Printing Office. Washington, D.C. 317 pp.
Ripper, W. E. & George, M. (1965). *Cotton Pests of the Sudan.* Blackwells Scientific Publications, Oxford. 345 pp.
Singh, S. R., van Emden, H. F. & Taylor, T. A. (1978). *Pests of Grain Legumes: Ecology and Control*. Academic Press, London. 454 pp.
Tourte, R. & Fauché, J. (1954). La 'rosette' de l'Arachide. *Bull. agron. Fr.d'out. mer*., **13**, 155.

Way, M. J. (1953). The relationship between certain ant species with particular reference to biological control of the coreid, *Theraptus* sp. *Bull. ent. Res.*, **44(4)**, 667–91.

Way, M. J. (1976). Diversity and stability concepts in relation to tropical insect pest management. *Proc. 9th Ann. Conf. and 10th Anniv. Celeb. ent. Soc. Nig.*, 68–93.

Williams, C. B. (1947). The fields of research in preventive entomology. *Ann. appl. Biol.*, **34(2)**, 175–85.

7 Plant Resistance

7.1 What is plant resistance?

Painter (1951) defined plant resistance to insects as the 'amount of heritable qualities' possessed by the plant which influences the ultimate degree of damage done by the insect. Earlier, Snelling (1941) used a slightly different definition and considered resistance as 'including those characteristics which enable a plant to avoid, tolerate, or recover from the attacks of insects under conditions that would cause greater injury to other plants of the same species'. Beck (1965) defines plant resistance as 'being the collective heritable characteristics by which a plant species, race, clone or individual may reduce the probability of successful utilization of that plant as a host by an insect species, race, biotype, or individual'. The term resistance is relative and is definable only in terms of other varieties of a species: plant resistance represents the inherent ability of a crop variety to restrict, retard or overcome pest infestations. For our purposes resistant varieties of plants are less damaged or less infested by the pest than other varieties in the field under comparable environmental conditions and stage of growth. Plant resistance to insects may be expressed at various stages of the insect/host plant relationship. Nine plant reactions affecting arrival, colonization and development of insect populations have been described (van Emden & Way, 1972; Dodd & van Emden, 1979). Resistance can be measured in terms of plant infestation level, damage inflicted, crop loss, etc.

7.2 Degree of resistance

In practical terms resistance is really the level of damage from a pest, ranging from zero damage through reduction of crop yield to death of the host plant. Plant resistance to insects follows a graded series from little restraint on successful colonization by insects to plant immunity (van Emden, 1966; van Emden & Way, 1973). Painter (1951) has classified the degree of resistance in terms of infestation level and degree of damage in the following manner:

(1) *Immunity*: a variety that cannot be infested or injured at all by a specific insect species under any known conditions. Anything less than immunity is 'resistance'.
(2) *Highly resistant*: varieties which suffer little damage by a specific insect under a given set of conditions.
(3) *Low level of resistance*: varieties of a species which are damaged less by a pest than the average (cultivated varieties) damage for the crop.
(4) *Susceptible*: varieties which show average or more than average damage by an insect, reverse of resistant.
(5) *Highly susceptible*: varieties which are readily infested and suffer considerably more damage than average damage by an insect pest under consideration.

Other terms include moderately resistant and tolerant. Once a plant has been infected it is said to be tolerant if it continues to grow and bear a crop (though there may be a reduction in yield) despite the infection.

Dahms (1972*a*) identified sixteen possible criteria which can be used for evaluation of insect resistance in crop plants. A degree rating system is usually used, for example, 0 (highly resistant) to 5 (highly susceptible). The following criteria are the most useful in screening a large number of plant varieties:

(*a*) Visual damage rating of infested plant varieties i.e. the damage which is visible externally.
(*b*) Determination at regular intervals of the number of plants that survive infestation.
(*c*) Determination of yield loss by comparison of infested and non-infested plots.

The choice of criteria would depend on the plant and the insect species involved.

7.3 Phenomena related to resistance

Painter (1951) has pointed out that certain phenomena may enable the plants to escape pest attack, but that does not mean that the plants are resistant once attacked. He has used the term pseudo-resistance for such phenomena. The phenomena included in this category are:

(1) *Host evasion* Early maturing varieties may avoid pest attack by early maturity.
(2) *Induced resistance* Some environmental factors may temporarily confer increased resistance on the plant. These include changes in the nutrients available to the plant.
(3) *Escape by chance* Among a group of infested plants some remain uninfested. This may be due to mere chance, e.g. incomplete infestation. Only the studies of the progenies of such plants would establish if the plant is resistant when infested.

7.4 Mechanism of resistance

Painter (1951, 1958*a*) divided host-plant resistance into three major types: (*i*) tolerance; (*ii*) antibiosis; (*iii*) preference and non-preference. Escape due to some characteristic of the variety, rather than by chance, is another mechanism of resistance. As pointed out earlier, Painter (1951, 1958*a*) labelled this mechanism 'host evasion' and considered it a kind of pseudo-resistance. Feeny (1976) included this sort of escape in a category which he termed 'non-apparency'. This type of resistance is termed as 'escape resistance' in this book.

The above terms may be considered as follows:

Tolerance Beck (1965) omitted tolerance as 'a resistant type insect-plant relationship' but retained the terms preference and antibiosis. Tolerance is the term used when the resistant plant is capable of supporting a population of insects without loss of vigour. The tolerance component of resistance involves the plant more than the insect in the insect-plant interaction. Painter (1968) considered tolerance to be present when the plant is able to reproduce well despite an insect population equal to that which damages susceptible hosts. Such plants have the vigour to withstand the pest attack and grow well despite a heavy infestation and possess the ability to repair rapidly the damage caused by the pest and also replace at least some of the destroyed organs. Horber (1980) notes that tolerance is 'often confused with low resistance'. Pathak & Saxena (1979) believe that the ability of tolerant plants to survive infestations for a longer period permits a longer exposure of the insects to their natural enemies. These authors state that tolerance is *frequently* valuable in integrated pest-control programmes.

Antibiosis This term is employed when resistant plants cause adverse effects on the biology (e.g. survival, development and reproduction) of the insect. Mortality of early instars has been cited frequently as evidence of antibiosis (Dahms, 1972*b*). Sometimes the resistant plant contains a level of some nutrients that is too low to support the insects which therefore cannot live and reproduce on the plant. Presence of either a feeding deterrent or a larval growth inhibitor or a combination of these would confer resistance to pest attack. Stem-borer infestation in rice plants may be influenced by the plant which in resistant varieties adversely affects larval growth. Similarly, rice varieties resistant to green leaf hopper attack either possess toxic materials or are unable to provide for the nutrition of the insects (Pathak, 1969).

Preference and non-preference In this type of resistance the plant characters affect the behaviour of the insect during orientation for food, shelter and oviposition. A plant may have special features which makes it less attractive or unacceptable to the pest. Recently, Kogan & Ortman (1978) have suggested the term 'antixenosis' to replace non-preference. It conveys the idea that the plant is avoided as a 'bad host' by the insect (Kogan and Ortman, 1978. Horber, 1980). Cuthbert & Davis (1972) working on factors contributing to cowpea-*Curculio* resistance in Southern Peas, found that the preference of the adult beetle *Chalcodermus aeneus*, for pods of particular lines of Southern Pea and the success of adults in penetrating the pods accounted for most of the differences between resistant and susceptible lines. Pathak & Saxena (1979) state that in field plantings, non-preferred crop varieties frequently escape infestation and even when insects are caged on non-preferred hosts, they tend to lay fewer eggs and produce smaller populations than on susceptible hosts. Roberts *et al.* (1979 have shown that leaf pubescence is effective in reducing natural populations of the hessian fly, *Mayetiola destructor*, on wheat. Webster (1975) has provided a useful bibliography on the association of plant pubescence and insect resistance. A range of physical and chemical mechanisms developed by plants to resist insect attack has been summarized by Southwood (1973) and Norris & Kogan (1980).

Escape resistance (= host evasion = non-apparency in part). Evidence of the past few years points to at least certain host evasion tactics being heritable and they often influence the degree of damage done by insects. Feeny (1976) has dwelt extensively on 'non-apparency' and includes in this category asynchrony of the pest and the susceptible stage of the host. Resistance of some cultivars of sweet potato, *Ipomoea batatas,* to the sweet potato weevil, *Cylas formicarius elegantulus*, is due to escape which Barlow & Rolston (1981) have shown to be some attribute of the sweet potato that is neither antibiosis nor non-preference. A variety of factors are known to contribute to the resistance of a plant, and escape resistance would appear to deserve a fresh approach when studying the basis of host plant resistance.

Multiple mechanisms Investigations indicate that a multiplicity of mechanisms influence the ultimate degree of plant damage by insect pests. The resistance of a plant may be due to more than one category of resistance (e.g. Painter, 1958*b*; Norris & Kogan, 1980). According to Gallun (1972), antibiosis is an important adjunct to non-preference in the resistance of plants to insects because dissimilar mechanisms of resistance coupled to a resistant variety may retard the development of insect biotypes capable of overcoming resistance. According to Pathak & Saxena (1979), both antibiosis and non-preference may influence or even disrupt one or more of the following stages of an insect's establishment on a plant:

(*i*) orientation; (*ii*) feeding; (*iii*) metabolic utilization of food; (*iv*) growth; (*v*) survival and egg production; (*vi*) oviposition and (*vii*) hatching of eggs.

Beck (1965), Chapman & Bernays (1977) and Norris & Kogan (1980) etc. have stressed the importance of studying the basic cause of resistance and there have been many investigations, especially those concerned with identifying the chemical factors associated with host-plant resistance. These factors are related to the presence or absence of special substances and differing amounts of nutrients affecting various gustatory responses of the insect (see Beck, 1965; Kennedy, 1965 for reviews of this subject). There is considerable evidence that insects respond positively to various plant nutrients including sugars, amino acids, polypeptides and vitamins. The nutrients may act as feeding stimulants or as co-factors and synergists to more specific stimulants and can play a dominant role in host selection, feeding and colonization by various insect species. Chalfant & Gaines (1973), in their studies on correlation between chemical composition of the Southern Pea and varietal resistance, reported a positive significant correlation between concentration of total carbohydrates in the hull and feeding punctures in the hull and pea. Haskell & Mordue (1969) mention that the nutritional requirements of various locusts suggest that naturally-occurring compounds act as phagostimulants which are also nutritionally essential to the locusts. Maxwell (1972) has reviewed the relationship of nutritional factors to resistance, particularly as it relates to antibiosis, preference, and overall pest management. Actually, genetic manipulation of the nutritional quality of insect's host would appear to be an impractical approach for the future. According to Maxwell (1977*a*) a plant developed 'so that it may be nutritionally inadequate for the insect might also be equally inadequate for man or the animals that would utilize the crop'. Earlier, Feeny (1976) expressed the view that nutritional deficiencies can be corrected for, after harvesting, either by making up deficient ingredients or by combining complementary foods in the diet.

7.5 Factors affecting resistance

Optimum level of resistance varies with the crop as well as the pest. It may depend on their location and would tend to vary with the economic returns of the crop. However, the following factors are known to affect the expression and stability of resistance.

7.5.1 Biochemical and morphological factors

Some biochemical and morphological factors affecting the resistance of plants were briefly discussed earlier. Biochemical factors, to a large extent, affect the behaviour and metabolic processes of the pest while morphological factors mostly influence the mechanisms of locomotion, feeding, oviposition, ingestion and digestion of the pest. Staedler (1977) reviewed insect–plant interactions in the light of studies based on electrophysiological recordings from sensory organs and other means, and provided useful references to morphological factors affecting resistance of plants to insect pests. Among morphological factors, colour and shape of the plant have been associated with long range perception of host plants (see for example, Stephens, 1957; Dunn & Kempton, 1976, Norris & Kogan, 1980 etc.). Cell wall thickness has been observed to interfere with feeding and oviposition mechanisms of insects (see, for example, Patanakamjorn & Pathak, 1967; Blum, 1968; Cuthbert & Davis, 1972; Norris & Kogan, 1980 etc.). Calcium and silica deposits in the epidermal wall of some members of Graminae have been shown to be responsible for the resistance to certain pests (see, for example, Miller *et al.*, 1960; Blum, 1968 etc.). Resistance to some stem-borers has been attributed to the stem solidness (see, for example, Howe, 1949; Norris & Kogan 1980 etc.). The effect of pubescence of a plant surface on the insect varies greatly but usually interferes with the feeding, ingestion, oviposition and locomotion of an insect (see Webster, 1975 for a bibliography of the subject as well as Norris & Kogan, 1980).

Recently, interest in elucidation of some of the chemical aspects of resistance has greatly increased (Maxwell, 1977*a*). However, while considerable success has been achieved in breeding crops resistant to insect pests, not enough is known about the chemistry of resistance in such cases (Maxwell, 1977*a*). Available knowledge suggests that chemically-based resistance is a 'major component of the plant's total defense armament against herbivores' (Norris & Kogan, 1980). Identification of the chemical basis of host-plant resistance greatly improved the efficiency of programmes for breeding resistant varieties (Kogan, 1977). Some examples of chemicals known to influence host plant resistance are as follows (see Norris & Kogan, 1980 for full treatment):

(1) Gossypol (1 1,6,6,7,7-hexahydroxy-5-5- diisopropyl-3,3-dimethyl-2,2-binaphthalene)-8,8-dicarboxaldehyde), a polyphenolic yellow pigment on cotton plants has been

demonstrated as a source of resistance to the bollworm *Heliothis zea*, and the tobacco budworm *H. virescens*. Cotton with a high gossypol content in the flower buds (2–3 times the amount found in cultivated varieties) were found to inhibit larval growth and suppressed populations of *H. virescens* (Lukefahr & Houghtaling, 1969). The feasibility of increasing the content of gossypol and other pigments of cotton plants with a view to protect the plant from pest attacks has been suggested by authors such as Lukefahr *et al.* (1966, 1971). Already the cultivated varieties of *Gossypium barbadense* containing a high level of gossypol are known to be almost immune to bollworm attack. It has also been shown (Singh & Weaver, 1972) that boll weevil, *Anthonomus grandis*, shows non-preference for strains of cotton containing a high level of gossypol. Efforts have been made to transfer the high-gossypol character into good agronomic cottons (Sappenfield *et al.*, 1974) and varieties that carry the character are now available (Maxwell, 1977*b*).

(2) Phenolic compounds in plants have been implicated in host defence against virus infections (Kosuge, 1969), fungal disease (Cruickshank & Perrin, 1964), and in pine appear to be involved in resistance to wasp attacks (Glenn *et al.*, 1971).

(3) The resistance of *Zea mays* to first-brood European corn borer, *Ostrinia nubilalis* (Lepidoptera) has been found to be due to the presence of a chemical DIMBOA (2,4-dihyroxy-7- -methoxy-1,4 benzoxazin-3-one) (Klun *et al.*, 1967) formed as a result of enzymatic conversion of a glucoside present in an uninjured corn plant. The importance of deterrent compounds released enzymically during tissue damage by feeding is now being appreciated. Woodhead & Bernays (1978) report that *Sorghum bicolor* is best protected from attack when very young, by the release on biting, of HCN from the non-deterrent cyanogenic glucoside, dhurrin. An enzyme system, probably present in the cytoplasm of the plant, is responsible for the hydrolysis of dhurrin.

(4) Oryzanone (*p*-methylacetophenone), a chemical extract from rice, has been found to attract ovipositing females of the striped stem-borer, *Chilo suppressalis*, which laid more eggs on a treated than on an untreated surface (Munakata &

Okamoto, 1967). Experiments have shown that resistant rice varieties containing this substance had lower dead heart incidence than those of other varieties (Pathak & Saxena, 1979).

(5) The resistance of lucerne to its pests has been attributed to higher saponin content, especially medicagenic acid, predominant in some resistant varieties.

In their discussion of the biochemical and morphological bases of resistance, Norris & Kogan (1980) state 'seldom, if ever, is one factor responsible for the resistance observed in a plant. This is particularly unlikely if a complex of pests is involved. The picture is incomplete, but a few examples show that several factors interact in the process'. These authors cite the case of the rice stem-borer, *Chilo suppressalis* where resistance to this pest in the rice plant has been shown to result from a combination of leaf blades with a hairy upper surface, tight leaf sheath wrapping, small stems with ridged surfaces, and thicker hypodermal layers (Patanakamjorn & Pathak, 1967). Similarly resistance in cotton to the pink bollworm, *Pectinophora gossypiella*, has been found to involve lack of bracts, glabrous leaves, cell proliferation, high gossypol content, and nectarless character (Agarwal *et al.*, 1976).

7.5.2. *Environmental factors*

Environmental factors such as temperature, light, relative humidity, soil fertility and soil moisture are known to affect the ability of plants to resist pest attacks. Thus, a variety that exhibits resistance in one locality or environment may be susceptible in another. Painter (1951) discussed some of the environmental factors in detail. Horber (1974) briefly touched on the subject but Tingey & Singh (1980) have recently reviewed the subject in detail.

Earlier reference was made to induced resistance. Environmental factors influence fundamental physiological processes of the plant as well as the pest and these may interact to make plant resistant or non-resistant temporarily. Generally at increased and decreased temperatures there is a loss of resistance. Similarly there is loss of resistance at decreased light intensity and at increased relative humidity. Plants grown under different levels of soil fertility tend to behave differently towards insect pests, the biology of which may in turn be affected.

Healthy crops are known to invite more pests, but the crops may be able to withstand the attacks by mere vigour. At least in some crops, high nitrogen fertilizer application tends to increase pest infestation. Soil moisture, e.g. good or excessive irrigation, may affect the character of resistance, but too little work exists on this aspect to make any safe generalizations. The resistance induced by environmental factors, as stated earlier, is not controlled by genes and hence is temporary in nature. However, as Tingey & Singh (1980) conclude, a knowledge of environmental influences on plant resistance may enable us to use induced resistance as a 'adjunct to genetic resistance in management of arthropod pests'. These authors point to the possibility of using induced resistance as 'the option of external manipulation at desired magnitudes and intervals'.

7.5.3 Genetic factors

To be valuable in a breeding programme, qualities of resistance possessed by a plant must be heritable. Two types of inheritance of resistance are generally recognized.

(1) Horizontal or general resistance: this type of resistance is governed by many minor genes, i.e. it is polygenic. In this type of resistance crosses are made within several varieties having low levels of resistance with the expectation of increasing resistance by accumulating allelles for resistance at several loci. Resistance terms (horizontal and vertical resistance) are extensively used in plant pathology (see Robinson 1971, and Van der Plank, 1968 for details).

(2) Vertical or specific resistance: this type of resistance is governed by a major gene (monogenic) rather than a change in gene frequency. With such resistance, gene to gene (Robinson, 1973) relationship exists between hosts and pests. The horizontal resistance approach is strongly favoured by some authors (Robinson, 1976).

Whatever the type of resistance, its genetics is studied from crosses of resistant and susceptible parents. Several extensive studies exist on the inheritance of resistance and different gene actions have been found in different studies on plant resistance.

For example, in their studies on the genetics of resistance to rice insects, Athwal & Pathak (1972) report that inheritance of resistance to brown and green hoppers in several rice varieties was controlled by a single dominant gene for resistance. In a review of the genetic factors affecting expression and stability of resistance, Gallun & Khush (1980) state that major genes generally impart high levels of resistance. But for every major gene found in the host there is a corresponding gene for virulence in the insect. Thus there is a greater chance of an insect biotype being selected where resistant varieties are grown widely. Polygenic resistance is 'biotype- non-specific' and is considered more stable. Both types of resistance are important in the crop improvement programmes (Gallun & Khush, 1980). Pathak & Saxena (1979) believe that with a better understanding of the factors affecting resistance in plants and their inheritance, genetic manipulation of resistance will become much more effective.

7.6 Development of resistant varieties

For the development of plant varieties resistant to pest attack, as stated earlier, there must exist genes for resistance which can be combined with commercially desirable qualities. For this purpose, diversified germ plasm is required. This usually means the screening of hundreds of varieties and thousands of breeding lines of a particular species or a plant for resistance to pests. It may require searching for varieties where the plant has originated as well as a study of related species or genera (Harlan & Starks, 1980). If the genetic base of resistance is narrow, its stability may be threatened due to development of insect biotypes capable of overcoming this resistance. For this reason it has been suggested that every effort should be made to use genes for general resistance and tolerance obtained from diverse sources and add to them genes for specific resistance (Browning, 1974).

From the work of many investigators over the past 10 years, it has become abundantly clear that the study of plant resistance is a complex operation and usually requires co-operation of workers from different scientific disciplines. Starks & Doggett (1970) note that the development of maize varieties resistant to European corn borer, *Ostrinia nubilalis*, required the following preliminary work:

(1) Study of the biology and behaviour of the insect and its relationship to the host plant;

(2) Development of techniques for obtaining a uniformly high level of infestation in the field. For this it may be necessary to use laboratory-reared insects. Latest advances in this field include the development of artificial diets for the continuous rearing of insect pests and screening the cultivars for confirmation of resistance.

(3) Obtaining simple, rapid, practical but reliable criteria for measuring resistance. These must be suitable for statistical analysis.

(4) Identification of the basis of resistance. For example, in the case of maize, chemicals such as DIMBOA linked to *Ostrinia* in the U.S.A., anatomical features, such as hard stem, loose leaf sheaths, hairiness etc. may be important.

In Ghana, scientists at the Cocoa Research Institute, Tafo have directed considerable effort to developing varieties resistant to swollen shoot virus which is transmitted by mealybugs. The complexity of the situation is indicated by the scheme set out in Fig. 4. It will be apparent that co-operation from the disciplines of entomology, mycology, agronomy and physiology is required in the process. At the moment there is no method of control of the swollen shoot disease available, other than destroying the infected trees. Development of resistant cocoa varieties seems to provide the best and only solution to the problem. However, in developing resistance to one pest, it is important to guard against the development of biotypes or susceptibility to other pests. The development of multiple resistance, high yield potential and consumer acceptance is the ultimate aim. Elsewhere in Africa, resistant cultivars are being developed for cowpea at IITA (International Institute of Tropical Agriculture, Ibadan, Nigeria) where Singh (1978) reports a collection of over 10 000 cowpea accessions with the plant breeders adding several thousand segregating lines every year. This has enabled resistance to field pests such as leafhopper (*Empoasca dolichi*) and pod borer (*Maruca testulalis*) to be studied extensively. Progress with the identification of pest and disease resistance in cowpea plants has been discussed by Singh (1980) and Singh & Allen (1980). Resistant cultivars of rice are also being screened by scientists of the West African Rice

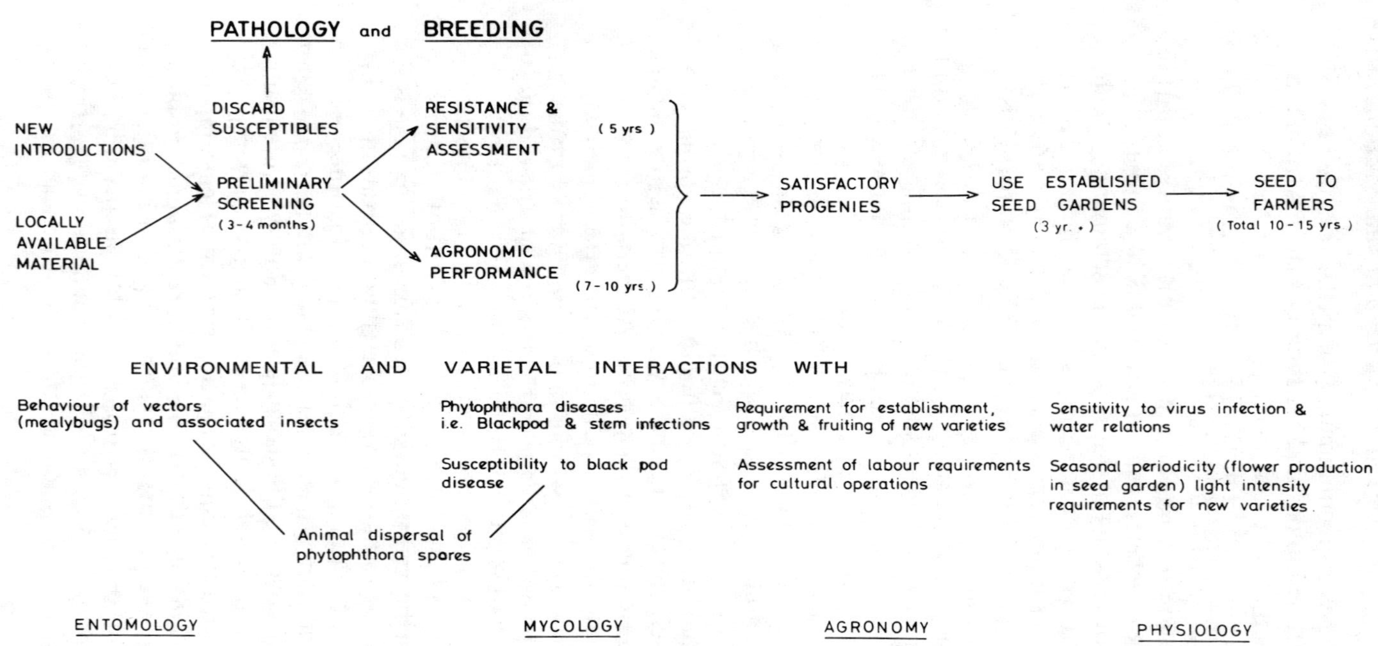

Figure 4 Scheme for assessing progeny for resistance to cocoa swollen shoot virus (after Legg, 1972; reproduced by permission of the Controller, Her Majesty's Stationery Office).

Development Association. Bidaux (1978) has reviewed the screening for horizontal resistance to rice blast in Africa.

7.7 Resistant varieties and pest control

Crop resistance, from the point of view of the farmer, is perhaps the easiest, most economical and effective way of controlling insect pests and diseases. It requires fewer technological inputs, creates no environmental hazards and is generally compatible with other methods of pest control (Pathak & Saxena, 1976). The use of American rootstocks resistant to phylloxera, *Phylloxera vitifoliae*, discovered about 1870, is still the major means of control of this pest of European grape varieties (*Vitis vinifera*). Resistance of such high level is very dramatic and highly useful in keeping the pest populations at low levels. The use of resistant varieties has reduced the hessian fly and wheat stem sawfly from a serious problem to the status of minor pests and averted an annual crop loss of 10 million dollars in the United States alone.

Resistant varieties are known to effect a cumulative reduction in pest numbers when compared to susceptible varieties (Pathak & Saxena, 1976) and form a key component in any pest management programmes. As pointed out by Horber (1972), 'resistant varieties may improve the effectiveness of insecticides and make it possible to omit or reduce treatments, and thus escape or lessen undesirable residues or side effects'. Table 3 shows the effect of plant resistance on the use of insecticide to control insects pests on cowpeas. A number of recent reviews e.g. Jepson (1954), and Mathes & Charpentier (1969) on varieties of sugar-cane and control of lepidopterous stalk-borer populations; Horber (1974) on insect resistance in forage legumes; Gallun *et al.* (1975) on insects attacking cereals; Bottrell & Adkisson (1977) on cotton insect pest management; Pathak (1969) and Cheng (1977) on resistant rice varieties to insect pests; Singh (1980) and Singh & Allen (1980) on resistance to cowpea pests; Pathak & Saxena (1979) and Adkisson & Dyck (1980) on a number of selected crops, all point to the enormous value of resistant varieties in pest management programmes. Maxwell *et al.* (1972) in a detailed review of more significant papers on plant resistance conclude that 'there have been numerous outstanding successes in

Table 3 Insect pests on cowpeas; affect of plant resistance on control with insecticides (after IITA, 1978, slightly modified).

Type of cowpea	Resistance status	Yield (kg/ha) With insect control	Without insect control
ER-1	Resistant due to early maturity	866	689
VITA-5	Moderately resistant	557	313
TVx 930-0IB	Moderately resistant	1001	444
Local	Susceptible	433	126

breeding for resistance. The value of these resistant crops cannot be fairly estimated except to say that it would range in the high millions of dollars annually'. Bottrell (1979) has stated that most serious agricultural pests in the United States have been controlled, at least in part of their range, by the use of resistant crop varieties. Savings as a result of their propagation run into billions of dollars annually.

Plant resistance is available at virtually no cost to the farmers yet it helps to stabilize yields. Farmers can divert their resources to other desirable inputs. Resistant varieties also spread rapidly without much extension effort (Dyck, 1974). Perhaps the most attractive feature of using pest resistant varieties, for crops grown in developing countries, as noted by Adkisson & Dyck (1980), is that virtually no skill in pest control or cash investment is required of the grower. Moreover, it appears to be cheaper to develop resistant varieties than new pesticides. In some cases, partial resistance may be all that is required. Actually, slight plant resistance is thought to be valuable as it imposes less selection pressure which, at least in some crops, may otherwise lead to the development of insect biotypes tolerant to the original resistance.

7.8 Literature cited

Agarwal, R. A., Banerjee, S. K., Singh, M. & Katiyar, K. N. (1976). Resistance to insects in cotton. II. To pink bollworm, *Pectinophora gossypiella* (Saunders). *Cotton Fibres Trop.*, **31**, 217–21.

Adkisson, P. L. & Dyck, V. A. (1980). Resistant varieties in pest management systems. pp. 233–51. In *Breeding Plants Resistant to Insects.* Eds F. G. Maxwell and P. R. Jennings. John Wiley and Sons, New York. 683 pp.

Athwal, D. S. & Pathak, M. D. (1972). Genetics of resistance to rice insects. In Rice Breeding. *Int. Rice Res. Inst., Los Baños, Philippines.* pp. 375–86.

Barlow, T. & Rolston, L. H. (1981). Types of host plant resistance to the sweet potato weevil found in sweet potato roots. *J. Kansas Entomol. Soc.*, **54(3)**, 649–57.

Beck, S. D. (1965). Resistance of plants to insects. *Ann. Rev. Ent.*, **10**, 207–32.

Bidaux, J. M. (1978). Screening for horizontal resistance to rice blast (*Pyricularia oryzae*) in Africa. pp. 159–74. In *Rice in Africa*. Eds I. W. Buddenhagen and G. J. Persley. Academic Press, London. 356 pp.

Blum, A. (1968). Anatomical phenomena in seedlings of sorghum varieties resistant to the sorghum shoot fly, *Atherigona varia soccata. Crop Sci.*, **8**, 388–90.

Bottrell, D. R. (1979). *Integrated Pest Management*. Council on Environmental Quality. U.S. Government Printing Office, Washington, D.C. 120 pp.

Bottrell, D. G. & Adkisson, P. L. (1977). Cotton insect pest management. *Ann. Rev. Ent.*, **22**, 451–81.

Browning, J. A. (1974). Relevance of knowledge about natural ecosystems to development of pest management programs for agro-ecosystems. *Proc. Amer. Phytopathol. Soc.*, **1**, 191–9.

Cheng, C. H. (1977). The possible role of resistant rice varieties in rice brown plant hopper control. pp. 214–29. In the *Rice Brown Planthopper*, compiled by Food and Fertilizer Technology Centre for the Asian and Pacific Region, Taiwan. 258 pp.

Chalfant, R. B. & Gaines T. P. (1973). Cowpea curculio: correlation between chemical composition of the Southern pea and varietal resistance. *J. Econ. Ent.*, **66(5)**, 1011–13.

Chapman, R. F. & Bernays, E. A. (1977). The chemical resistance of plants to insect attack. *Pontificiae Academiae Scientiarum Scripta Varia,* **41**, 603–43. (Published as: Marini-Bettolo, G. B. (ed.) (1977). Natural products and the protection of plants. Proceedings of a study week at the Pontifical Academy of Sciences. October 18–23, 1976. Elsevier Scientific Publishing Company, Amsterdam. 846 pp.)

Cruickshank, I. A. M. & Perrin, D. R. (1964). Pathological function of phenolic compounds in plants, pp. 511–44. In *Biochemistry of Phenolic Compounds.* Ed. J. B. Harborne. Academic Press, London and New York. 618 pp.

Cuthbert, F. P. & Davis, B. W. (1972). Factors contributing to cowpea curculio resistance in Southern peas. *J. Econ. Ent.*, **65(3)**, 778–81.

Dahms, R. G. (1972*a*). Techniques in the evaluation and development of host plant resistance. *J. Environ. Quality,* **1**, 254–9.

Dahms, R. G. (1972*b*). The role of host plant resistance in integrated insect control, pp. 152–67. In *Control of Sorghum Shootfly.* Eds M. G. Jotwani and W. R. Youngs. Oxford and IBH. Publ. Co., New Delhi. 324 pp.

Dodd, G. D. & van Emden, H. F. (1979). Shifts in host plant resistance to the cabbage aphid (*Brevicoryne brassicae*) exhibited by Brussels sprout plants. *Ann. appl. Biol.*, **91**, 251–62.

Dunn, J. A. & Kempton, D. P. H. (1976). Varietal differences in the susceptibility of Brussel sprouts to lepidopterous pests. *Ann. appl. Biol.*, **82**, 11–19.

Dyck, V. A. (1974). Insect pest management in rice: principles and practices. *Pesticides Annual* 1974 (India), pp. 69–71.

van Emden, H. F. (1966). Plant insect relationships and pest control. *Wld. Rev. Pest Control*, **5**, 115–23.

van Emden, H. F. & Way, M. J. (1973). Host plants in the population dynamics of insects, pp. 181–99. In *Insect/Plant Relationships*. Ed. H. F. van Emden, Symposia of the Royal Entomological Society of London: Number 6. Blackwell Scientific Publications, Oxford. 215 pp.

Feeny, P. (1976). Plant apparency and chemical defense. pp. 1–40. In *Biochemical Interactions between Plants and Insects. Recent Adv. Phytochem*. Vol. **10**. Eds J. W. Wallace and R. L. Mansell. Plenum Press, New York. 425 pp.

Gallun, R. L. (1972). Genetic interrelationship between host plant and insects. *J. Environ. Quality*, **1**, 259–65.

Gallun, R. L. & Khush, G. S. (1980). Genetic factors affecting expression and stability of resistance, pp. 63–85. In *Breeding Plants Resistant to Insects.* Eds F. G. Maxwell and P. R. Jennings. John Wiley and Sons, New York. 683 pp.

Gallun, R. L., Starks, K. J. & Guthrie, W. D. (1975). Plant resistance to insects attacking cereals. *Ann. Rev. Ent.*, **20**, 337–57.

Glenn, W. T., Getahm, A. & Cress, D. C. (1971). Resistance in Barley to the Greenbug, *Schizaphis graminum*. 1. Toxicity of phenolic and flavonid compounds and related substances. *Ann. Ent. Soc. Amer.*, **64(3)**, 718–22.

Harlan, J. R. & Starks, K. J. (1980). Germplasm resources and needs. pp. 253–73. In *Breeding Plants Resistant to Insects.* Eds F. G. Maxwell and P. R. Jennings. John Wiley and Sons, New York. 683 pp.

Haskell, P. T. & Mordue (Luntz), A. J. (1969). The role of the mouthpart receptors in the feeding behaviour of *Schistocerca gregaria. Entomologia experimentalis et applicata*, **12**, 591–610.

Horber, E. (1972). Plant resistance to insects. *Agric. Sc. Rev.*, 1–10, 18.

Horber, E. (1974). Techniques, accomplishments and potential of insect resistance in forage legumes, pp. 312–42. In *Proceedings of the Summer Institute on Biological Control of Plant Insects and Diseases.* Eds F. G. Maxwell and F. A. Harris. University Press of Mississippi, Jackson. 647 pp.

Horber, E. (1980). Types and classification of resistance, pp. 15–21. In *Breeding plants resistant to insects*. Eds F. G. Maxwell and P. R. Jennings. John Wiley and Sons, New York. 683 pp.

Howe, W. L. (1949). Factors affecting the resistance of certain cucurbits to the squash borer. *J. Econ. Ent.*, **42**, 321–6.

IITA (International Institute of Tropical Agriculture) (1978). Research Highlights, 1977. 72 pp.

Jepson, W. F. (1954). A critical review of the world literature on the lepidopterous stalk borers of tropical graminaceous crops. *Commonw. Inst. Ent. London*. 127 pp.

Kennedy, F. S. (1965). Mechanisms of host plant selection. *Ann. appl. Biol.*, **56**, 317–22.

Klun, J. A., Tipton, C. L. & Brindley, T. A. (1967). 2,4-Dihydroxy-7-methoxy-1, 4-benzoxazin-3-one (DIMBOA), an active agent in the resistance of maize to the European corn borer. *J. Econ. Ent.*, **60(6)**, 1529–33.

Kogan, M. (1977). The role of chemical factors in insect/plant relationships. *Proc. XV. International Congr. Ent. Washington D.C.* 211–27.

Kogan, M. & Ortman, E. E. (1978). Antixenosis – a new term proposed to define Painter's 'non-preference' modality of resistance. *Bull. Ent. Soc. Amer.*, **24(2)**, 175–6.

Kosuge, T. (1969). The role of phenolics in host response to infection. *Ann. Rev. Phytopath.*, **7**, 195–222.

Legg, J. T. (1972). Measures to control spread of cocoa swollen shoot disease in Ghana. *PANS*, **18(1)**, 57–60.

Lukefahr, M. J., Houghtaling, J. E. (1969).Resistance of cotton strains with high gossypol content to *Heliothis* spp. *J. Econ. Ent.*, **64**, 588–91.

Lukefahr, M. J., Houghtaling, J. E. & Graham, H. N. (1971). Suppression of *Heliothis* populations with glabrous cotton strains. *J. Econ. Ent.*, **64**, 486–8.

Lukefahr, M. J., Noble, W. & Houghtaling, J. E. (1966). Growth and infestation of bollworms and other insects on glanded and glandless strains of cotton. *J. Econ. Ent.*, **59**, 817–20.

Mathes, R. & Charpentier, L. J. (1969). Varietal resistance in sugarcane to stalk moth borers, pp. 175–88. In *Pests of Sugarcane.* Eds J. R. Williams, J. R. Metcalfe, R. W. Mungomery and R. Mathes. Elsevier Publishing Company, Amsterdam, 568 pp.

Maxwell, F. G. (1972). Host plant resistance to insects – nutritional and pest management relationships, pp. 599–609. In *Insect and Mite Nutrition*. Ed. J. G. Rodriguez. North-Holland Publishing Company, Amsterdam. 702 pp.

Maxwell, F. G. (1977*a*). Host-plant resistance to insects-chemical relationships, pp. 299–304. In *Chemical Control of Insect Behaviour.* Eds H. H. Shorey and J. J. Jr. McKelvey. Wiley, New York. 512 pp.

Maxwell, F. G. (1977*b*). Plant resistance to cotton insects. *Bull. Ent. Soc. Amer.*, **23**, 199–203.

Maxwell, F. G., Jenkins, J. N. & Parrott, W. L. (1972). Resistance of plants to insects. *Adv. Agron.*, **24**, 187–265.

Miller, B. S., Robinson, R. J., Johnson, J. A. & Pannoya, B. W. K. (1960). Studies on the relation between silica in wheat plants and resistance to Hessian fly attack. *J. Econ. Ent.*, **53**, 995–9.

Munakata, K. & Okamoto, D. (1967). Varietal resistance to rice stem borers in Japan, pp. 419–30. In *The Major Insect Pests of the Rice Plant.* Proceedings of a Symposium at the International Rice Research Institute September 1964. Johns Hopkins Press, Baltimore, Maryland. 729 pp.

Norris, D. M. & Kogan, M. (1980). Biochemical and morphological bases of resistance, pp. 23–61. In *Breeding Plants Resistant to Insects.* Eds F. G. Maxwell and P. R. Jennings. John Wiley and Sons, New York. 683 pp.

Painter, R. H. (1951). *Insect Resistance in Crop Plants.* MacMillan, New York. 520 pp.

Painter, R. H. (1958*a*). Resistance of plants to insects. *Ann. Rev. Ent.*, **3**, 267–90.

Painter, R. H. (1958*b*). The study of resistance to aphids in crop plants. *Proc. 10th International Congr. Ent. Montreal,* Vol. **3**, 1956 (1958), 451–8.

Painter, R. H. (1968). Crops that resist insects provide a way to increase world food supply. *Kans. Agri. Exp. Stat. Bull.*, **520**, 1–22.

Pathak, M. D. (1969). Stemborer and leafhopper-planthopper resistance in rice varieties. *Entomologia experimentalis et applicata*, **12**, 789–800.

Pathak, M. D. & Saxena, R. C. (1976). Insect resistance in crop plants. *Current Adv. Sci.*, **27**, 1233–52.

Pathak, M. D. & Saxena, R. C. (1979). Insect resistance, pp. 270–92. In *Plant Breeding Perspectives.* Eds J. Sneep and A. J. T. Hendriksen. Centre for Agricultural Publishing and Documentation, Wageningen. 460 pp.

Patanakamjorn, S. & Pathak, M. D. (1967). Varietal resistance of rice to the Asiatic rice borer, *Chilo suppressalis* (Lepidoptera : Crambidae), and its association with various plant characters. *Ann. Ent. Soc. Amer.*, **60**, 287–92.

Roberts, J. J., Gallun, R. L., Patterson, F. L. & Foster, J. E. (1979). Effects of wheat leaf pubescence on the Hessian fly. *J. Econ. Ent.*, **72(2)**, 211–14.

Robinson, R. A. (1971). Vertical resistance. *Rev. Plant Path.*, **50**, 233–9.

Robinson, R. A. (1973). Horizontal resistance. *Rev. Plant Path.*, **52**, 483–501.

Robinson, R. A. (1976). *Plant Pathosystems.* Springer-Verlag, Berlin. 184 pp.

Sappenfield, W. P., Stokes, I. G. & Harrendorf, K. (1974). Selecting plants with high square gossypol. *Proceedings of the Beltwide Cotton Production Research Conference*, **27**, 87–93.

Singh, I. D. & Weaver, J. B. Jr. (1972). Growth and infestation of bollweevils of normal-glanded, glandless, and high-gossypol strains of cotton. *J. Econ. Ent.*, **65(3)**, 821–4.

Singh, S. R. (1978). Resistance to pests of cowpea in Nigeria, pp. 267–79. In *Pests of Grain Legumes: ecology and control.* Eds S. R. Singh, H. F. van Emden and T. A. Taylor. Academic Press, London. 454 pp.

Singh, S. R. (1980). Biology of cowpea pests and potential for host plant resistance, pp. 398–421. In *Biology and Breeding for Resistance to Arthropods and Pathogens in Agricultural Plants*. Ed. M. K. Harris. Texas A. & M University Bulletin. MP – 1451. 605 pp.

Singh, S. R. & Allen, D. J. (1980). Pests, diseases, and protection in cowpeas, pp. 419–43. In *Advances in Legume Science*. Eds R. Summerfield and H. Bunting. Ministry of Agr. For. Fish. Her Majesty's Govt. and Kew Gardens. 667 pp.

Snelling, R. O. (1941). Resistance of plants to insect attack. *Bot. Res.*, 7, 543–86.

Southwood, T. R. E. (1973). The insect/plant relationship – an evolutionary perspective. *Symp. Roy. Entomol. Soc. Lond.*, **6**, 3–30.

Staedler, E. (1977). Sensory aspects of insects plant interactions. *Proc. XV. International Congr. Ent. Washington D.C.* 228–48.

Starks, K. J. & Doggett, H. (1970). Resistance to a spotted stem borer in Sorghum and Maize. *J. Econ. Ent.*, **63(6)**, 1790–5.

Stephens, S. G. (1957). Source of resistance of cotton strains to the boll weevil and their possible utilization. *J. Econ. Ent.*, **50**, 415–18.

Tingey, W. M. & Singh, S. R. (1980). Environmental factors influencing the magnitude and expression of resistance, p. 87–113. In *Breeding Plants Resistant to Insects.* Eds. F. E. Maxwell and P. R. Jennings. John Wiley & Sons, New York. 683 pp.

Van der Plank (1968). *Disease Resistance in Plants.* Academic Press, New York. 206 pp.

Webster, J. A. (1975). Annotated bibliography: Association of plant hairs and insect resistance. USDA –ARS. 1297. 18 pp.

Woodhead, S. & Bernays, E. A. (1978). The chemical basis of resistance of *Sorghum bicolor* to attack by *Locusta migratoria. Entomologia experimentalis et applicata*, **24**, 123–44.

8 Biological Control

Although in recent years biological control is often used to include any biologically-based methods of pest suppression, in its traditional sense it means the manipulation of natural enemies of pests to reduce their populations to a level where economic losses due to them are tolerable. The natural enemies are often abundant and prevent a species from reaching pest proportions especially in long established communities such as forests. Natural enemies may be vertebrates or invertebrates but are usually parasites, predators or pathogens of insect pests. In nature, the abundance of a species is largely governed by its relationship with other organisms – competitors and natural enemies and a 'balance of nature' which tends to keep the numbers of any species within limits is usually maintained. However, as discussed earlier, modern agriculture often upsets balances which have been established over long periods. Modern methods of biological control attempt to restore this balance in a number of ways (see later).

8.1 Biological control agents

These agents may be classified as follows:

(1) Parasites and predators
(2) Phytophagous insects
(3) Dung beetles
(4) Insect pathogens such as viruses, fungi, protozoa and nematodes

8.1.1 Parasites and predators

Insects important as biological control agents may be parasites and predators. Parasitic insects, often called parasitoids, lay their eggs in or on the bodies of other arthropods and develop at the expense of the host, eventually killing it. Most belong to the orders Hymenoptera and Diptera. Most species of the hymenopteran superfamilies Ichneumonoidea (Fig. 5a) and Chalcidoidea (Fig. 5b) are either parasites or hyperparasites (parasites of parasites), attacking the eggs, larvae, pupae, or very

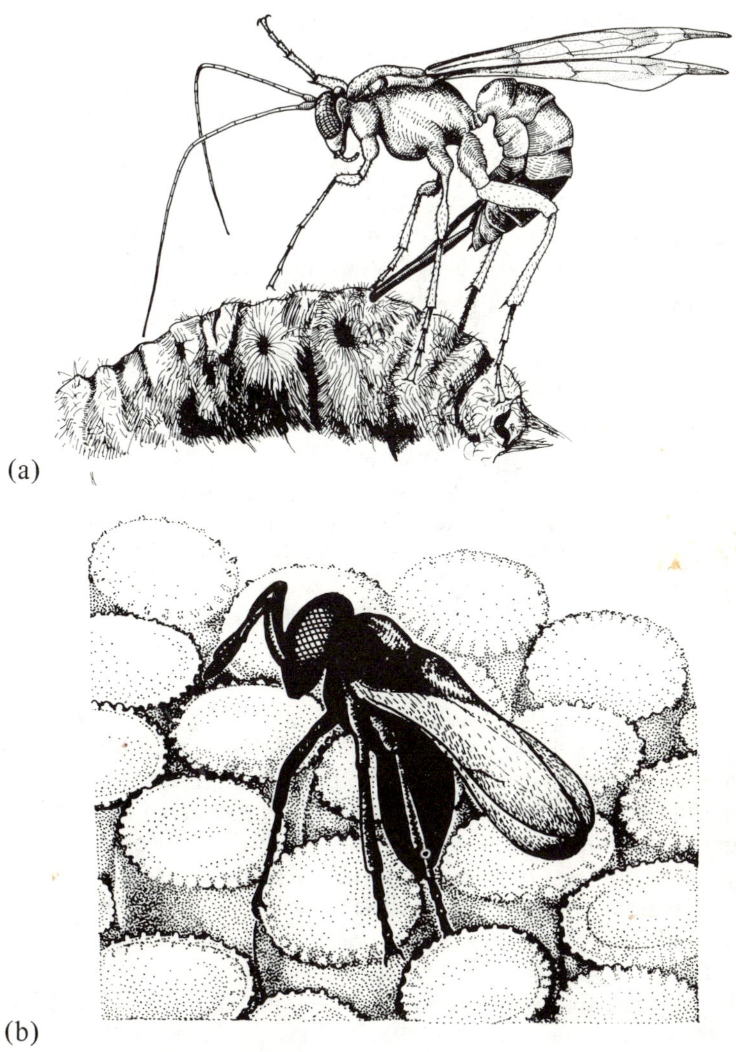

Figure 5 (**a**) Ichneumonid parasite laying an egg in a moth pupa. (**b**) Scelionid parasite ovipositing in pentatomid eggs.

rarely the adult. The family Braconidae (Ichneumonoidea) alone contains over 1000 genera, and one genus, *Apanteles*, contains over 1150 species (see Askew, 1971; Shervis & Shenefelt, 1973 for useful references). Members of the dipteran family, Tachinidae, are uniformly parasites of insects or other invertebrates. Parasites are often host specific, which enhances their effectiveness in the control of particular pest species.

Predators kill their prey by direct attack and an individual may consume many prey individuals. Syrphidae (Diptera), Reduviidae (Heteroptera), Chrysopidae and Hemerobiidae (Neuroptera), Dermaptera, Mantodea (Dictyoptera), Coccinellidae, Cantharidae and Carabidae (Coleoptera) are examples of groups containing predacious insects. Predators are usually larger than their victims and tend to be polyphagous. They may have chewing or sucking mouthparts, may hunt their prey on the ground, on vegetation or in flight, or may trap it by various means. Certain species exhibit division of labour between the non-predacious mother who locates and sometimes hunts the prey, and the larvae who actually feed upon it. The components of arthropod predator – prey interaction have been reviewed by Hassell (1978) who provides many references to earlier work. Kiritani & Dempster (1973) discuss different approaches to the quantitative evaluation of natural enemy effectiveness (see later) and provide a useful bibliography on the subject.

The idea of using insect enemies to control pests, as noted by Norris (1968), DeBach (1974) and others, was known in certain parts of the Orient as early as the twelfth century. Here farmers *augmented* natural populations of enemies (ants) by putting ant nests in fruit trees. For centuries the Chinese have maintained colonies or purchased colonies of the predator ant, (*Oecophylla smaragdina*) and placed them in orange trees to reduce leaf-feeding insect populations. One of the best known recent examples of classical biological control relates to cotton cushony scale (*Icerya purchasi*) which caused havoc in the citrus orchards of California. A beetle (*Rodolia cardinalis*) (Fig. 6) which preys upon the scale was introduced from Australia in 1888. This inoculative release of the natural enemy has since kept the scale under effective check not only in California, but many other parts of the world. One

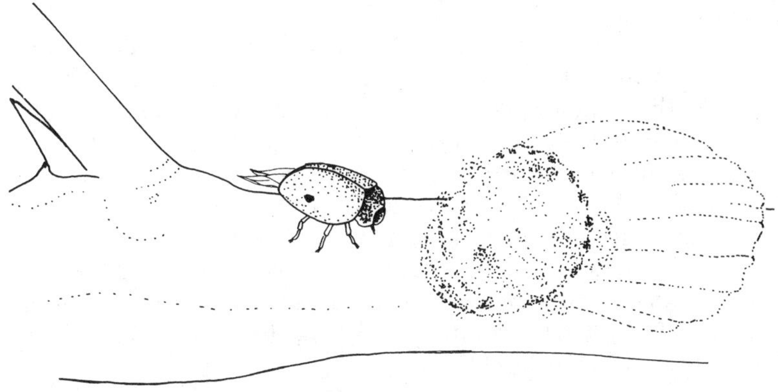

Figure 6 The predatory beetle, *Rodolia cardinalis* which preys on the cotton cushony scale, *Icerya purchasi.*

hundred and seventy-one beneficial insect species are reported to have been imported into the United States during the past 85 years (Sailer, 1976). These deliberately introduced natural enemies have established themselves on about 55 foreign insect pests and 5 weeds. Waterhouse & Wilson (1968) have provided some interesting examples from Australia of pest control by the introduction of natural insect enemies. Throughout the world, up to January, 1975, Huffaker (1975) states that there were 213 cases of partial to complete success involving the introduction of natural enemies of important pest insects and arthropod relatives, snails, and weeds where they had not previously occurred.

On the African scene, perhaps the most successful example of the biological control programme has been the control of Kenya coffee mealybug, (*Planococcus kenyae*) accidentally introduced into Kenya from Uganda. After about 15 years of unsuccessful parasite and predator introductions from various parts of the world, and mistaken identity of the pest, the control was eventually achieved by a parastitic wasp, *Anagyrus* sp. near *kivuensis* introduced from the natural home of the mealybug, i.e. Uganda. Another highly successful African example is the control of eucalyptus snout beetle (*Gonipterus scutellatus*) in South Africa by the introduction of an egg parasite, *Patasson nitens*, from South Australia. The parasite dispersed rapidly and effectively controlled the pest in most parts of that country (Tooke, 1955).

Table 4 summarizes some of the biological control attempts in Africa. For details of various projects, reference should be made to the work of Greathead (1971). Simmonds (1972) has dealt generally with biological control in the tropics. Current status of biological control of forest pests in Nigeria has been reviewed by Akanbi (1978) while Kumar (1979) has discussed the possibilities of biological control of cocoa pests in West Africa. For detailed studies on natural enemies of some West African cocoa pests references should be made to Firempong & Kumar (1975), Owusu-Manu (1976), Akotoye & Kumar (1977) and Kumar & Nutsugah (1979).

8.1.2 *Phytophagous insects*

Control of weeds by means of phytophagous insects is now a well established practice. One of the best known examples is the control of prickly pear (*Opuntia inermis* and *O. stricta*), of American origin, which occupied some 4 000 000 ha of land in Queensland, Australia in 1900 and 24 000 000 ha by 1925. Half of the area was so densely infested that it was useless for agriculture. The weed was brought under control by the introduction of a moth, *Cactoblastis cactorum* (Fig. 7) from South America. The caterpillars of the moth fed voraciously on fleshy cladodes of prickly pear and brought it under control within a few years. Many other highly successful examples of the control of weeds by the introduction of insect pests are documented in literature.

Spectacular example of weed control by phytophagous insects in Africa is that of the prickly pear, *Opuntia ficus-indica*, introduced into South Africa from Mexico. The weed spread widely over a period of 200 years and despite herbicidal and mechanical control had covered an area of about 900 000 ha by 1942. The cochineal insect, *Dactylopius opuntiae*, introduced from Queensland (material originally came from Texas and Arizona) in 1937, was mainly responsible for the collapse of nearly 750 000 ha of prickly pear jungle (Annecke & Moran, 1978; Moran, 1981). The action of the cochineal insects was greatly enhanced through the use of DDT which was used to

Figure 7 Caterpillars of the moth, *Cactoblastis cactorum*, used in the biological control of the prickly pear cacti.

protect them from coccinellid predators. Cochineal insects have been considered to be very important in South Africa, in the control of another notorious weed, the jointed cactus, *Opunitia aurantica* (Moran & Annecke, 1979). It is interesting to note that the pyralid moth, *Cactoblastis cactorum*, which was such an outstanding success against the smaller prickly pears, *O. inermis* and *O.stricta* in Australia, played only a minor role in control of *O. ficus-indica* in S. Africa. Elsewhere in Africa initial success in establishing *Paulinia* on the aquatic weed *Salvinia* on Lake Kariba in Zimbabwe has also been reported (Andres & Bennett, 1975). In West Africa, two obnoxious weeds, *Eupatorium odoratum* (Hall *et al.,* 1972, Schröder, 1970; Sheldrick, 1968) and *Lantana camara* (Schröder, 1970), are currently engaging the attention of biological control workers. Worldwide reviews of biological control of weeds are by Huffaker (1959), DeBach (1964) and Wilson (1964).

Table 4 Biological control attempts in Africa (after Greathead, 1971, modified) – record of natural enemies sent from one country to another in the Ethiopian region. *Denotes introduced species of natural enemies.

Host	Natural enemy	From	To	Date	Result
Pest					
ORTHOPTERA: ACRIDIDAE					
Nomadacris septemfasciata	*Wohlfahrtia euvittata* (Sarcophagidae)	S. Africa	Mauritius	1933	Not released
HEMIPTERA: PENTATOMIDAE					
Antestiopsis spp.	*Corioxenos antestiae* (Strepsiptera)	Tanzania	Kenya	1936	?
	Bogosia rubens (Tachinidae)	Uganda	Tanzania	1965, 1967	Recovered
Nezara viridula	*Asolcus basalis* (Scelionidae)	S. Africa	Zimbabwe	1956	Already present
COREIDAE					
Pseudotheraptus wayi	*Ooencyrtus* sp.	Kenya	Zanzibar	1959	?
	Ooencyrtus sp. (Encyrtidae)	Zanzibar	Kenya	1959	?
TROPIDUCHIDAE					
Numicia viridis	**Tytthus mundulus*	Mauritius	S. Africa	1966–9	?
	**T. parviceps* (Miridae)				?
APHIDIDAE					
Eriosoma lanigerum	**Aphelinus mali* (Aphelinidae)	S. Africa	Zimbabwe	1961	Established

Table 4 (*continued*)

	Exochomus melanocephalus (Coccinellidae)	S. Africa	Ascension	1900	?
(Aphids)	*Adalia flavomaculata* (Coccinellidae)	S. Africa	Ascension	1902	?
	Chilomenes lunata (Coccinellidae)	S. Africa	Ascension	1902	?
	Oenopia cinctella (Coccinellidae)	S. Africa	Ascension	1902	?
ALEURODIDAE					
Aleurocanthus woglumi	**Eretmocerus serius* (Eulophidae)	Seychelles	Kenya	1958	Established
MARGARODIDAE					
Icerya purchasi	**Rodolia cardinalis*	S. Africa	Kenya	1917	Not established
	(Coccinellidae)	S. Africa	St. Helena	1896–8	Established
Icerya seychellarum	**Rodolia cardinalis*	Mauritius	Seychelles	1930	Established
	(Coccinellidae)	S. Africa	Mauritius	1915, 1937–8	Established
	Cryptochetum monophlebi	Madagascar	Mauritius	1952–3	Established
	(Cryptochetidae)	Mauritius	S. Africa	1959	Not released
		Madagascar	S. Africa	1961	
ORTHEZIIDAE					
Orthezia insignis	**Hyperaspis jocosa* (Coccinellidae)	Kenya	Malawi	1959	? Not established

Table 4 (*continued*)

Host	Natural enemy	From	To	Date	Result
PSEUDOCOCCIDAE					
Dysmicoccus brevipes	**Cryptolaemus montrouzieri* (Coccinellidae)	S. Africa	Mauritius	1939–40	Established
Planococcus kenyae	**Cryptolaemus montrouzieri* (Coccinellidae)	S. Africa	Kenya	1929–30	Established
	Anagyrus kivuensis etc. (Encyrtidae)	Congo	Kenya	1937	?
	Schizobremia coffeae (Cecidomyiidae)	Tanzania	Kenya	1934	?
	parasites and predators (various)	Uganda	Kenya	1927, 1935 1938, 1951	Some established
Planococcoides njalensis	**Anagyrus* sp. nr. *kivuensis* (Encyrtidae)	Kenya	Ghana	1948	Not established
Rastrococcus iceryoides	**Rodolia cardinalis* (Coccinellidae)	S. Africa	Tanzania	1955	?
DIASPIDIDAE					
Aonidiella aurantii	**Comperiella bifasciata* (Encyrtidae)	S. Africa	Swaziland	1969	?
Aspidiotus destructor	**Cryptognatha nodiceps* (Coccinellidae)	Principé	Angola	1960–3	Established
	Pseudomicrocera henningsi (Fungi)	Sierra Leone	Seychelles	1929	Already present
Aulacaspis tegalensis	*Chilocorus discoideus*	Uganda	Mauritius	1969	?
	C. distigma (Coccinellidae)	Kenya	Mauritius	1969	?
	Physcus sp. (Aphelinidae)	Uganda	Mauritius	1969	?

Table 4 (*continued*)

Parlatoria blanchardii	*Chilocorus distigma* (Coccinellidae)	Senegal	Mauritania (via France)	1967–8	?
(Coconut Coccidoids)	*Chilocorus distigma*	E. Africa	Seychelles	1936	Established
	C. wahlbergi	E. Africa	Seychelles	1936	Not established
	Exochomus ventralis	E. Africa	Seychelles	1936	Established
	E. flavipes (Coccinellidae)	E. Africa	Seychelles	1936	Established
	Aphytis chrysomphali (Aphelinidae)	E. Africa	Seychelles	1936	Not established
	**C. nigritus*	Mauritius	Chagos	1956	?
	Scymnus sp.	Mauritius	Chagos	1956	?
(Coccidoids)	**C. nigritus*	Mauritius	Agelega	1955	?
	**Lindorus lophanthae* (Coccinellidae)				
LEPIDOPTERA: GELECHIIDAE					
Phthorimaea operculella	**Apanteles subandinus* (Braconidae)	S. Africa	Zimbabwe	1969	?
	**Copidosoma uruguayensis*	S. Africa	Zimbabwe	?	?
	(Encyrtidae)	S. Africa	Zambia	1968	?
	**C. uruguayensis*	Mauritius	Madagascar	1967–8	?
		Mauritius	Seychelles	1966	Not recovered
	Nythobia stellenboschensis (Ichneumonidae)	S. Africa	Seychelles	1968	Not recovered
Pectinophora gossypiella	*Bracon kirkpatricki* (Braconidae)	Kenya	Sudan	1928	Already present

Table 4 (*continued*)

NYMPHALIDAE					
Papilio demodocus	*Carcelia evolans* (Tachinidae)	Madagascar	Mauritius	1965	? Not released
		Madagascar	Réunion	1965	Established
	Brachmeria cowani (Chalcididae)	Madagascar	Réunion	1965	?
DIPTERA: CULICIDAE					
Anopheles gambiae	*Coelomomyces* sp. (Blastocladiales)	Zambia	Kenya	1963	Not established
Anopheles spp.	*Pachypanchax playfairi* (Cyprinodontidae)	Seychelles	Zanzibar	?1912	'Success'
		Seychelles	Tanzania	1920	?
	'*Panchax* spp.'	?	Kenya	1937	?
	Poecilia reticulata (Poeciliidae)	Kenya	Uganda	1940	Established
	'*Carpes maillort*'	Réunion	Madagascar	1914	?
TEPHRITIDAE					
Ceratitis capitata	Parasites	W. Africa	S. Africa	1913	? Released
Dacus spp.	*Opius phaeostigma*	S. Africa	Mauritius	1934	Dead on arrival
(Fruit Flies)	*Dirhinus giffardii* (Chalcididae)	W. Africa (via Hawaii)	Mauritius	1957–63	Not established
		(Mauritius)	Réunion	*ca.* 1960	Not established
MUSCIDAE					
Glossina spp.	*Splangia glossinae* (Pteromalidae)	Tanzania	Nigeria	1925	Already present
Musca domestica	*parasites	Mauritius	Seychelles	1966	Not recovered

Table 4 (*continued*)

Host	Natural enemy	From	To	Date	Result
PSYCHIDAE					
Kotochalia junodi	parasites of Psychidae	Madagascar	S. Africa	1934, 1936	Not released
PYRALIDAE					
Chilo sacchariphagus	*Pediobius furvus* (Eulophidae)	Uganda	Mauritius	1969	?
	Xanthopimpla citrina	Mauritius	Réunion	1953, 1960	Established
	**X. stemmator* (Ichneumonidae)	Mauritius	Réunion	1953, 1960	Established
	Apanteles flavipes (Braconidae)	Mauritius	Madagascar	1955, 1960–1	Established
	**X. stemmator*	Mauritius	Madagascar	1958	Not recovered
	**Diatraeophaga striatalis*	Madagascar	Mauritius	1964	? Not established
		Madagascar	Réunion	1964, 67, 68	? Established
NOCTUIDAE					
Heliothis armigera	*Trichogramma lutea* (strain) (Trichogrammatidae)	Zimbabwe	S. Africa	? 1936	Established
Sesamia calamistis	*Apanteles sesamiae* (Braconidae)	Kenya	Mauritius	1952	Established
		Mauritius	Réunion	1953–5	Established
		Mauritius	Madagascar	1955, 1967	Not established
		Uganda	Madagascar	1968	Not established
	Pediobius furvus (Eulopidae)	Uganda	Madagascar	1968	Established

Table 4 (*continued*)

Host	Natural enemy	From	To	Date	Result
COLEOPTRA: SCARABAEIDAE					
Clemora smithi	Scoliids	Madagascar	Mauritius	1917, 1932, 1934, 1936–6	Some established
	Tachinids	Madagascar	Mauritius	1936	Not established
	Scoliids	S. Africa	Mauritius	1938–9	Not established
	Tachinids	S. Africa	Mauritius	1939–40	Not established
	Adapsilia latipennis (Pyrgotidae)	S. Africa	Mauritius	1938–40	Not established
Oryctes monoceros	*Scolia ruficornis* (Scoliidae)	Zanzibar	Seychelles	1949–51	Established
	Platymeris laevicollis (Reduviidae)	Zanzibar (via Pakistan)	Seychelles	1963–4	Not recovered
	Neochryopus savagei (Carabidae)	Nigeria	Seychelles	1960	Not recovered
Oryctes rhinoceros	**Scolia oryctophaga*	Mauritius	Chagos	1960	?
	S. ruficornis (Scoliidae)	Kenya	Mauritius	1966	?
	Platymeris laevicollis (Reduviidae)	Zanzibar	Mauritius	1962–7	? Not established
		(also via Pakistan)	Chagos	1963	?
	Scarites madagascariensis (Carabidae)	Madagascar	Mauritius	1965–6	?
Oryctes tarandus	*Scolia oryctophaga* (Scoliidae)	Madagascar	Mauritius	1917	Established

Table 4 (*continued*)

Schizonycha sp.	**Campsomeris erythrogaster*	Mauritius	Kenya	1951	Probably not est.
	**C. phalerata*	Mauritius	Kenya	1951	Probably not est.
	**C. lachesis*	Mauritius	Kenya	1951	Probably not est.
	**Scolia carnifex* (Scoliidae)	Mauritius	Kenya	1951	Probably not est.
	**Tiphia parallela* (Tiphiidae)	Mauritius	Kenya	1951	Probably not est.
CURCULIONIDAE					
Gonipterus scutellatus	**Patasson nitens* (Mymaridae)	S. Africa	Kenya	1945	Established
		S. Africa	Mauritius	1946	Established
		Kenya	Mauritius	1946	Established
		Mauritius	Madagascar	1948	Established
		S. Africa	St. Helena	1958	Established
SCOLYTIDAE					
Hypothenemus hampei	*Prorops nasuta* (Bethylidae)	Uganda	Kenya	1930	Already present
ACARI: TETRANYCHIDAE					
Tetranychus marianae	*Stethorus jejeunus* (Coccinellidae)	S. Africa	Mauritius	1950	?
PULMONATA: ACHATINIDAE					
Achatina fulica	*Gonaxis quadrilateralis* (Streptaxidae)	Kenya	Seychelles	1958	Established
		Kenya (via Hawaii)	Mauritius	1961	Established

Table 4 (*continued*)

Host	Natural enemy	From	To	Date	Results
		Mauritius	Madagascar	? 1967	?
	**Euglandina rosea*	Mauritius	Seychelles	1966	?
	(Oleacinidae)	Mauritius	Madagascar	1962, 66, 68, 69	?
		Mauritius	Réunion	1966	?
HELICIDAE					
Helix aspersa etc.	*Gonaxis kibweziensis* (Streptaxidae)	Kenya	S. Africa	1957	Not established
	Gonaxis sp.	Kenya	S. Africa	1957	Not established
PLANORBIDAE					
Biomphalaria spp.	*Astatareochromius alluaudi*	Uganda	Tanzania	1960	Established
	(Cichlidae)	Uganda	Cameroun	ca 1960	?
RODENTIA:MURIDAE					
Rattus rattus	*Tyto alba* (Strigidae)	E. Africa	Seychelles	1949	Established
	owls	Madagascar	Mauritius	1901	Not established
WEEDS					
CACTACEAE					
Opuntia spp.	**Dactylopius ceylonicus*	S. Africa	Tanzania	1957	Established
	(Eriococcidae)	Tanzania	Kenya	1958	Established
		S. Africa	Mauritius	1914	Established
	**D. opuntiae*	Réunion	Madagascar	1923	Established

Table 4 (*continued*)

	**Cactoblastis cactorum* (Pyralidae)	S. Africa	Mauritius	1950	Established
VERBENACEAE					
Lantana camara	**Teleonemia scrupulosa* (Tingidae)	Kenya	Tanzania	1958	Established
		Kenya	Uganda	1960	Established
		Kenya	Zimbabwe	1961	Established
		Zimbabwe	Zambia	1962	Not established
		Mauritius	Madagascar	196-?	Established
	Hypena strigata (Noctuidae)	Kenya (via Hawaii)	S. Africa	1961-2	Already present

(a)

(b)

Figure 8 (**a**) Scarabaeid beetles at work in removing a patch of dung. (**b**) Scarabaeid beetles rolling the completed dung ball.

8.1.3 Dung beetles (*Scarabaeidae*)

Accumulation of dung has created serious problems in Australia since the introduction of cattle there. An individual animal produces about 10 pads daily. The rich, green vegetation surrounding a pad is avoided by all but starving cattle. Each animal may render unproductive 4% of a hectare of land per annum (Waterhouse & Wilson, 1968), resulting in a serious loss in improved pastures at high stocking rates. Additionally, cattle dung is the breeding ground of pest flies. Two types of beetles from Africa have been introduced into Australia to tackle the dung problem. One type attacks maggots in dung pads while the other breaks up the pads and buries them in tunnels under the soils thereby not only adding to soil fertility but also seriously interfering with the development of fly maggots. Africa·alone has more than 2000 species of dung beetles and some have been successfully introduced into Hawaii (Greathead, 1971). Dung beetles play a key role in removing the dung left by herds of mammals in Africa and elsewhere (Heinrich & Bartholomew, 1979) (Fig. 8).

8.1.4 Insect pathogens: viruses, bacteria, fungi, protozoa and nematodes

Insects are affected by diseases caused by viruses, fungi, protozoa and bacteria. Between 1500 and 2000 species of these organisms affecting insects have been described. Entomologists have long been aware of the potential of microbial control of pests. As early as the 1930s the European spruce sawfly (*Gilpinia hercyniae*), which caused massive destruction to spruce forests in Canada, was brought under effective control by a naturally-occurring virus accidentally introduced from Europe during imporation of its natural parasites. Tanda (1959), Burges & Hussey (1970), David (1975) and Tinsley (1979) collate most of the important literature concerning microbial control. Disease intensity among natural insect populations is largely a density-dependent factor of mortality. Thus disease, which appears to be relatively sparse in nature, often causes considerable mortality in populations of insects when mass-reared in the laboratory. The qualities to be considered in the microbial control agents are as follows:

(1) high virulence for target species in the field;
(2) harmlessness to non-target species including beneficial organisms and vertebrates;
(3) ease of production and storage for long periods without loss of virulence;
(4) capability of acting rapidly against the target species;
(5) resistance to harmful factors in the environment such as solar radiation, dessication, heat, changes in pH, etc.

Viruses Insects are known to be affected by seven types of viruses. They are distinguished in size and shape. To avoid the possibility of infection of animals and man, WHO and FAO have recommended that only one of these, baculoviruses, should be considered as pesticidal agents (see Summers *et al.*, 1975 for safety consideration of these viruses). These viruses have no properties in common with the viruses known to infest animals and plants, which are thus safe from the risk of accidental infection. Within the baculoviruses, work has been concentrated on nuclear polyhedrosis viruses (NPV) which are virulent, highly specific, are enclosed in a protective mass of protein which enables them, if not exposed to ultra-violet light, to survive in the environment for years. NPVs are known to attack a number of species of lepidopterous caterpillars. In the U.S.A., NPVs have been tried on a large scale against certain lepidopterous larvae in the genus *Heliothis*, including *H. zea* on cotton as well as *H. virescens*. After 12 years of experimentation no change was found in the specificity of the virus and no sign of any resistance on the part of the host. Field trials in Botswana (Roome, 1975) with an unpurified suspension of local *Heliothis armigera* nuclear polyhedrosis virus (NPV) against *H. armigera* larvae on sorghum were found to be as effective as a standard insecticide in curtailing losses of sorghum. This virus was also reasonably effective against *H. armigera* on cotton at very high rates of application. Viron H, the commercial *Heliothis zea* NPV preparation, was found to be less effective than the local virus (Roome, 1975). Two viral pesticides – Viron and VHZ have lately been put on the market but very recently these products were found to be much less effective in the field than under conditions of laboratory testing. The major drawbacks with the use of NPVs lie in the fact that they must be produced in living tissues of insects. The difficulty of obtaining large numbers of insects when required has been partly solved by the development of artificial diets on which to rear them. Another difficulty with virus applications for control is that they involve an incubation period of up to 10–20 days during which time the pest can cause considerable damage to the crop affected.

The available volume of information on viruses has rapidly increased in the last 35 years. However, much work remains to be done on the effects of viruses under field conditions. Often the available information seems contradictory, e.g. Allen *et al.* (1966) in U.S.A. carried out biological control of *Heliothis* species on cotton integrating a naturally-occurring predator-parasite complex and NPVs. These authors reported no adverse effects of the virus on the parasite. On the other hand Irabagon & Brooks (1974) working on the interaction of the parasite, *Campoletis sonorensis* and a NPV in the larvae of *H. virescens* found that the host–parasite–pathogen interactions are highly detrimental to the parasite. Further, NPVs are known to be inactivated by ultraviolet light and there is the added problem of the adverse effect of some leaf surfaces on virus. Thus a considerable amount of work needs to be done in developing suitable formulations for use under field conditions. Problems associated with the use of arthropod viruses in pest control have been reviewed by Falcon (1976) and Tinsley (1979). Many of the entomopathogens are excellent candidates for development into safe, effective microbial insecticides and have potential in the management of many insect pests.

Bacteria Spore-forming bacteria of the genus *Bacillus* have received considerable attention in the recent years as alternatives or adjuvants to insecticides for pest control. *Bacillus thuringiensis* has been tested against a wide spectrum of insects in the laboratory and the field. Most insects tested have been Lepidoptera or Diptera but species of Coleoptera, Hymenoptera, Orthoptera and Homoptera are also susceptible to the pathogen (Angus, 1968). *B. thuringiensis* has received particular attention due to its pathogenicity for larvae of Lepioptera (Heimpel, 1967). Over 150 species of lepidopterous larvae are known to be susceptible to *B. thuringiensis* which produces a proteinaceous parasporal body, an endotoxin, that is toxic to these larvae. Within minutes of exposure to preparations of *B. thuringiensis* gut paralysis occurs and this results in the cessation of feeding (see Somerville, 1977 for details of mode of action of the endotoxin). *B. thuringiensis* is extremely safe, having no adverse effects on beneficial insects, man, pets and other animals (Fisher & Rosmer, 1959). Commercial preparations (often termed biotic chemicals) of *B. thuringiensis*, e.g. Dipel, Thuricide, HPC, Biotrol and Bactospeine P.M., have been successfully used against several agricultural pests. For example, Mistric & Smith (1973) used it against tobacco bud worm, while Charpentier *et al.* (1973) found it to give highly effective control of sugar-cane borer, *Diatraea saccharalis*, in

Louisiana. Creighton & McFadden (1974) obtained effective control of the cabbage looper, *Trichoplusia ni*, and the cabbage worm, *Pieris rapae*, with sprays containing low level of Dipel and chlordimeform hydrochloride. Preparations of *B. thuringiensis* have also been used effectively for controlling lice infesting poultry (Hoffman & Gingrich, 1968) and larvae of the housefly in poultry manure by feeding the endotoxin to caged chickens (Burns *et al*., 1961). Some approved uses of this agent in the U.S.A. are given in Table 5. Very recently, the work of Morris (1977) indicates that the integrated approach using *B. thuringiensis* and chemical combinations in a viable alternative to the use of chemical pesticides alone to control spruce budworm in Canada.

In Ghana and Nigeria, a bacterial pathogen of *Sahlbergella singularis* gave encouraging results in both field and laboratory trials against cocoa pests, *Distantiella theobroma, S. singularis* and *Helopeltis bergrothi* but was ineffective against *Bryocoropsis laticollis* (Bolton, 1973). Trials in the field showed that mirid populations on treated cocoa trees were reduced by 67% in 96 hours whilst the control area showed a reduction of 35% in the same period. These experimental successes however, have not been followed up by large scale field trials. Taylor

Table 5 Approved uses of the insect disease agent, *Bacillus thuringiensis* (as indicated on commercial product labels) in the U.S.A. (after Bottrell, 1979).

Insect pests	Plants
Cabbage looper, diamondback moth, imported cabbage worm	Broccoli, cabbage, cauliflower, collards, kale, lettuce, mustard, spinach, turnip, greens
Cabbage looper	Beans, cucumbers, melons, potatoes
Cabbage looper, celery looper	Celery
Grape leaf folder	Grapes
Cabbage looper	Flowers (Chrysanthemums)
Bagworm, Douglas-fir tussock moth, elm spanworm, fall webworm, gypsy moth, red-humped caterpillar, spring and fall canker worms, tent caterpillar	Ornamental and shade trees

(1974) found Dipel a suitable alternative to insecticides for the control of a variety of lepidopterous larvae associated with okra, *Abelmseus esculentus*, in Nigeria. Lavabre *et al.* (1966) have demonstrated the usefulness of viral and bacterial pathogens in the control of caterpillars on cocoa in West Africa. It is difficult to understand why these studies have not been pursued. Mention should also be made here of failures with the use of *B. thuringiensis* preparations. Sutter (1969) showed that larvae and adults in corn rootworms, *Diabrotica* spp. are immune to attack by *B. thuringiensis* and *B. popillia*. Similarly, *B. thuringiensis* has not controlled bollworms on cotton. Lewis *et al.* (1974) in U.S.A., using *B. thuringiensis*, were unable to achieve the desired population reductions of the gypsy moth, *Porthetria dispar*. A further complication has been noted by Wilson & Burns (1968) who report that they were able to reduce resistance to the bacterium in houseflies (*Musca domestica*).

Fungi Numerous species of fungi attack and kill insects in nature. *Beauveria bassiana* causes muscardine disease of silkworm and is cosmopolitan in distribution. It seems to flourish in soil and humid environments close to the ground and has been recorded from a number of coleopterous hosts and their larvae. Tanda (1959) believes that the genus *Beauveria* offers most promise in microbial control because of its high pathogenicity for a large number of hosts. Field applications of *Beauveria* spp. have been found effective against a number of insect pests, e.g. Colorado potato beetle. Pathogenic fungi of the genus *Entomophthora* are well known to entomologists and plant pathologists. According to Angus (1977), species of this kind of fungus are important determinants of insect abundance, and in some cases are thought to be key factors in terminating insect outbreaks. In U.S.S.R., *Aspergillus ripens* has been applied to control the pentatomid, *Eurygaster integriceps*. Fungal epidemics on insects such as scales are occasionally encountered in nature. Evans (1974), in a review of the natural control of arthropods by fungi in the tropical high forest of Ghana suggests that pathogenic fungi may be important in the natural control of arthropod populations in humid tropical forests. Steinhaus (1949) also belives that entomogenous fungi in nature cause a regular and heavy mortality of many pests throughout the world. This view is supported by authors such as Rockwood (1950), Madelin (1960), and Le Pelley (1968) but not shared by Lack (1954) and Bucher (1964).

Fungi are directly subject to regulation by physical factors in the environment, requiring optimum conditions, especially proper humidity, to cause host mortality. They cannot therefore, be depended on for general field use. Workers have from time to time obtained mortality in the laboratory, presumably due to higher humidity in a closed system which tends to enhance the infectivity of the pathogen. Results under field conditions have usually been disappointing and when successful require repeated applications of massive doses to sustain control of insect pests. Thus far fungi have generally proved to be of little value as pest control agents.

Protozoa The most common protozoa infecting insects are those belonging to the subclasses Gregarinida, Coccidia and Microsporidia.

Of the Gregarinida, *Mattesia dispora* has been credited with field and laboratory mortality, in Europe and Australia of three lepidopterous pests of stored cereals and their products: *Ephestia cautella*, *E. kuehniella* and *Plodia interpunctella*. The microsporidian, *Nosema bombycis* causes the classic pebrine of the silkworm (Tanda, 1959). *Nosema* spp. have been tried against several lepidopterous caterpillar species. Angus (1977) reports that, in the U.S.A., barnbait formulations of *Nosema* sp., gave successful control of grasshoppers over 10 000 acres of rangeland.

Protozoa as insect pathogens are not highly virulent under field conditions and it is doubtful if they could ever be commercially marketed to control the pests. However, protozoa may affect the life processes of insects and thus make them susceptible to other forms of natural control. Recent developments indicate that some protozoans may be effectively used to control grasshoppers, boll weevils, citrus mites and mosquito larvae. It is obviously a field where much work remains to be done. According to Steinhaus (1967), 'the disappointing delay in the greater use of fungi and protozoa may be accounted for largely by the lack of basic research on these agents – basic research oriented towards such matters as their mass productions, selectivity of strains, environmental studies and field tests'.

Nematodes Research on entomogenous nematodes has revealed that a large number of these parasites attack living insects. In dissection of mirid and pentatomid bugs attacking cocoa in West Africa, nematodes are frequently found in the body cavity. Members of the genus *Neoaplectana* have been studied widely and in many cases the results of laboratory tests were promising but field trials proved disappointing. According to Simons (1978) there are possibilities of using neoaplectanids as biological control agents, especially those of *N. carpocapsae.* This nematode is said to be easily mass reared, stored and sprayed, it is non-susceptible to many insecticides, and can migrate swiftly in the soil (Simons, 1978). Angus (1977) reports that a species of meremithid is now available commercially (Skeeter Doom(R) for control of some species of mosquitoes).

From the works edited by Burges & Hussey (1971) it is clear that more research and development effort in the area of microbial control of insects is necessary before any great expansion in the use of microbial agents in pest control will occur.

8.2 The practice of biological control

For natural enemies to exist in an agro-ecosystem, a pest population must be present as well. Interactions between a pest and its natural enemies will result in an equilibrium which cannot be achieved by the enemy alone. This ensures further generations of prey/host and means we should be willing to tolerate sub-economic levels of damage by pests. Some insects, such as vectors of humans, animal and plant disease, are not well suited for control by natural enemies because the threshold for such pests is virtually zero. For example, the populations of *Planococcus citri*, the vector of swollen shoot virus disease of cocoa in West Africa are present at very low densities in the field. The low populations are the result of a rich fauna of their natural enemies but even a few pest individuals are sufficient to cause destruction of a large number of cocoa trees. Any attempt at control of such pests by use of natural enemies is unlikely to be successful and this indeed has been the experience of field work.

In practice the following three approaches to the use of natural enemies of pest management are available:

(*i*) importation, (*ii*) augmentation, and (*iii*) conservation.

Accurate identification of both the pest and its natural enemies is an essential prerequisite to success in biological control. Misidentification of the pest may misdirect the search for natural enemies which may result in waste of effort, or in neglect of search for potentially effective enemies. In Africa we are aware that the early history of the search for control of the Kenya coffee mealybugs, *Planococcus kenyae*, was bedevilled by confusion as to its identity (Greathead, 1971). It was not until its exact identity and likely place of origin was established that an effective parasite was introduced for its control. Compere (1961) reports that for nearly 50 years, attempts at biological control of the California red scale, *Aonidiella aurantii*, have been handicapped and confused by the failure to find any morphological differences between this scale and the closely-related yellow scale. Similar difficulties were encountered with their natural enemies. Compere (1961) states that 'probably no insect demonstrates better than the red scale the importance of sound classification and exact identification . . . ' (see Kumar, 1981 for the need for the correct identification of insect species in scientific work). For workers in the British Commonwealth, an identification service is operated by the Commonwealth Institute of Entomology, London and located at the British Museum of Natural History.

8.2.1 Importation

This is the classical approach in biological control and involves the introduction of suitable enemies of pests into an area where their host is a pest but they themselves do not occur. Under favourable conditions, beneficial species released in sufficient number will successfully colonize and become an integral part of the ecosystem. So far this technique has yielded by far the best results (see DeBach 1964, 1972 and Simmonds, 1967 for details) and has been largely, but not exclusively, directed against exotic insect pests. Such work however, involves a good deal of fundamental research on the control agent to be introduced. Generally a number of potential species from various parts of the world are screened. Extensive laboratory trials are conducted, firstly to determine the possible effectiveness of the predator/parasite/pathogen and secondly to make certain that it will not have other and more harmful side effects such as becoming a pest of plants or animals other than the target species. Guidelines for importation of new natural enemies mainly summarized by Sailer (1975), are as follows:

(1) Pests of foreign origin are the most promising subjects but indigenous pests with closely-related species occurring in other parts of the world are also likely choices.

(2) Chances of success are greatest against a crop with high economic threshold for damage or if the economic damage involves a non-marketable part of the crop.

(3) Introduction of a wide spectrum of enemies of the pest should be attempted with a view to obtain the best combination of enemies for a given situation.

(4) Search efforts should be directed in the original home of the pest or in climatically similar areas.

(5) A really effective natural enemy usually establishes itself quickly and spreads rapidly. If it does not it probably will not be fully effective even if it becomes established. Efforts should not be wasted on a species which fails to establish itself after 2–3 conducted attempts at introduction because the process of biological control is expensive and time-consuming.

(6) No geographical area of crop pest should be prejudged as unsuitable for introduction but the chances of success should influence the priorities.

(7) The following characteristics should be sought in natural enemies likely to be introduced:

   (*a*) high searching and dispersal ability; (*b*) high degree of host specificity, it is best if the enemy attacks only one host species or at most 2 to 3; (*c*) high fecundity and rapid development relative to the host; (*d*) good adaptation to new environment; (*e*) in the case of insect parasites and predators, it should not be phytophagous or prey on other parasites (hyperparasitism) of the species to be controlled; (*f*) in the case of pathogens, the organism must be harmless to all other living forms, animals and plants; and (*g*) pathogens should be resistant to harmful effects of environment (see section on viruses).

(8) The introduction programmes should be carried out by

trained personnel backed by adequate quarantine facilities. A project involving introduction of natural enemies from foreign lands should be undertaken by a well integrated team functioning as a single unit. This involves the following activities:

(*i*) selection of target species;
(*ii*) exploration, and discovery of natural enemies;
(*iii*) selection of candidates to be shipped;
(*iv*) propagation of stock, if possible, otherwise collection of material to be shipped. When possible, propagation will eliminate secondary parasites and may be less costly;
(*v*) quarantine, screening and rearing of selected material;
(*vi*) laboratory studies preliminary to release (see below); propagation if possible, if not continue importation.
(*vii*) field colonization;
(*viii*) recovery surveillance to determine establishment action to ensure dispersal throughout the area infested by host.
(*ix*) evaluation of results including cost/benefit analysis.

Source of natural enemies Before importing any control agents, the world distribution of the pest and its natural enemies should be determined. If the original home of the pest is known, this is the logical place to begin the search as here the possibility of finding an effective natural enemy is greatest (Andres & McMurty, 1979). Since climate determines the distribution of pests and their natural enemies, search for the latter, as said earlier, must include areas climatically similar to where releases are to be made. Dissimilarity between the climate of Europe and western United States has been frequently blamed for the limited success or failure of many introductions from Europe (Andres & McMurty, 1979).

Once a firm decision on the control agent and its areas of occurrence has been made, organizations such as the Commonwealth Institute of Biological Control (with headquarters in Trinidad) may be approached for assistance in securing the control agent. This institute has stations in several continents engaged in the search for promising natural enemies for use against specific pests. The United States Department of Agriculture (at Beltsville, Maryland) and Entomology Division of CSIRO, Canberra, Australia, also operate some stations in several overseas countries, searching for biological control agents. Alternatively, stock of an already known enemy may be obtained through arrangements with individuals or institutions. If one has the financial resources to visit the area of search, foreign cooperators can prove extremely valuable in locating the desired host plants and pest infestations.

Once sufficient stock of the required controlling agent has been collected, it must be properly packed to avoid damage through excessive heat or cold as well as escapes during air shipment. Parasites are usually best shipped in the pupal stage while predators are sent as immature stages or adults. On arrival, all insects must be cultured in quarantine, by providing them with suitable food, to establish a reproducing colony to make sure that no harmful organism such as hyper-parasites, diseases or other pests are associated with them. All pathogens must be tested to make sure that they are indeed the types being sought for propagation. In the case of the biological control of weeds, extensive tests with imported organisms are required to ensure that local plants of value will not be attacked. These studies must be satisfactorily completed before any release is made. Any agent that develops on a useful plant must be destroyed.

Colonization and propagation Stock of the species imported may be used to establish a laboratory colony. As noted by Sailer (1979), each species tends to be a special case as some species can be reared easily while others cannot be reared at all in the laboratory. The laboratory stock can be used to make periodic releases and agricultural officers and growers can assist in dissemination of the control agents. Colonization and application methods for viruses and bacteria require some further discussion. Insect viruses are obligate parasites and can grow only in living cells. Therefore, only intact cells or tissue cultures derived from intact insects are used for virus production (Heimpel, 1977). A number of workers have collected diseased larvae in large numbers, in the field, or used them to infect laboratory reared stock, reduced them to a fine powder in a blender and filtered to obtain a crude suspension. Such preparations are termed 'dirty' (Heimpel, 1977) as they tend to contain other micro-organisms such as protozoa or other viruses from the same host species. Some such preparations (see, for example, Daoust & Roome, 1974) have given very good results under field conditions. Field investigators in Africa working away from large laboratories can identify NPVs by making smears of the body contents of diseased larvae, staining them by the Giesma technique and examining under a microscope. Confirmation of identity can be subsequently obtained from institutions such as the N.E.R.C. Unit of Invertebrate Virology, Oxford, England.*

On a commercial scale the *Heliothis* virus for the commercial preparation 'Elcar' is produced in live *H. zea* larvae reared in the laboratory on an artificial diet, using as many as 200 000 larvae per day. At a suitable stage of larval development virus is introduced on the diet and the dead and dying larvae are harvested several days later. The virus is checked for any contaminants, bioassayed and formulated into a wettable powder. To protect the viruses after application, from harmful effects of sunlight, materials such as molasses have been used, sometimes as a protectant and as a sticker (Heimpel, 1977).

Japidemic® or Doom®, an effective bacterial preparation of the milky disease *Bacillus popilliae* of the Japanese beetle (*Popillia japonica*) is made from ground inoculated grubs of the pest and is mixed with an inert carrier such as chalk. It is applied to the soil where it is spread by rainwater, insects and other animals, thereby affecting the soil-dwelling grubs which are immature feeding stages of the Japaneses beetle (Bottrell, 1979). Nature's recycling of a natural control process is thereafter assured until the population of the grubs is controlled.

In contrast to Japidemic commercial preparations of *B. thuringiensis*, Biotrol®, Dipel® and Thuricide, are produced by fermentation techniques (Dulmage & Rhodes, 1971), a technique similar to that used to manufacture antibiotics. A few cells from a pure culture preserved in a test tube are sufficient to start the process. The bacteria are grown in a liquid medium in sealed 15 000 gallon tanks with appropriate aeration and

*A diagnostic manual for the identification of insect pathogens has been published by Poinar & Thomas (1978).

stirring. Molasses, corn steep liquor, proteins such as fish meal or soy flour, and minerals and vitamins are used to nourish the multiplying bacteria. The bacterium multiplies by elongating and dividing, and eventually oval-shaped and diamond-shaped crystals form within the ageing bacteria. These are released when the cells break down and the final fermentation product, containing vegetative cells, spores and crystals (toxins) is concentrated, formulated, assayed and then packaged for sale.

Application of pathogens Pathogens such as viruses and bacteria must be ingested in order to infect the target insect. This requires an absolute coverage of the plant surface (Heimpel, 1977). This author mentions a recommendation by Dr Falcon regarding the use of a spray unit that can deliver dropper sizes from 1–20 nm. Using ultra-low volume sprays with oil emulsion good results with bacteria and virus were reported. Although Heimpel (1977) expresses scepticism over the use of aeroplanes to provide coverage of pathogens in forest areas, Morris (1977), de Groot *et al*. (1979) and several other Canadian workers have reported the successful application of nuclear polyhedrosis virus using light aircraft over forest.

8.2.2 Augmentation

The autmentative approach may be inoculative or inundative. Inoculative release anticipates establishment and control of future generations of pests, inundative release on the other hand anticipates only control of the population and generation on which it is applied, with no expectation of long-term regulation. Sailer (1975) and Bottrell (1979) have provided a summary of the many efforts to apply the augmentive method in the field. Sailer (1975) also discusses the reason for failures in the use of this method in several situations.

The best known case of an augmentive approach is that of the inundative releases of the egg parasite, *Trichogramma*, a chalcid wasp. According to Dysart (1973), in 1969, 2.5 million hectares of the U.S.S.R. received releases of *Trichogramma* and efforts are now directed towards automated plants for mass production of the parasite. Similarly, during 1974 in Mexico, some 28 billion *Trichogramma* were produced and released against a number of lepidopterous pests (see Sailer, 1975 for references). In the United States, more than 3.5 billion *Trichogramma* are produced annually with more than 2 billion of these reared for commercial release. The results were reported to compare favourably with the reductions in pest populations achieved with insecticidal applications. Nguyen-Ban (1975) has reported success in rearing of *Trichogramma* on a semi-industrial scale in the Ivory Coast and their release in the cocoa-ecosystem for the control of *Earias biplaga*. In several countries, notably in U.S.A. (California), populations of mealybugs have been effectively kept in check by periodic releases of coccinellid beetles. Such releases may be required by the inability of the predator to survive adverse periods of weather in the wild. Releases of native parasites and predators to control insect pests biologically have taken place in several countries, especially the United States of America (Stinner, 1977, Ridgway & Vinson, 1977). The efficacy of inundative releases has been dealt with by Stinner (1977).

Recent work on the use of chemicals in modifying the behaviour and activity of natural enemies may provide a new weapon for the suppression of pest populations using biological control agents. For example, Lewis *et al.* (1972), found a 'host produced chemical' that greatly increases the oviposition activity of the *Trichogramma* females, thereby enhancing the degree of control from release of a given number of parasites. Response of parasitic insects to behavioural chemicals (kairomones) in locating hosts has been reviewed by Lewis *et al.* (1976). Hagen & Bishop (1979) in a discussion on increasing predation and parasitization through the use of supplementary foods and kairomones conclude, 'there is much promise in their use as additional tools to be employed by pest managers where naturally-occurring biological control agents require augmentation. They can be used in conjunction with releases of natural enemies'. For a full treatment of biological control by augmentation of natural enemies, reference should be made to the recent book by Ridgway & Vinson (1977).

8.2.3. Conservation

This involves creating situations favourable to colonization of crops by resident natural enemies. Effective manipulation of the environment in favour of natural enemies requires a good knowledge of the biology and behaviour of the pest as well as the enemy species. This may entail modified cultural practices such as interplanting of suitable crops, strip-planting as opposed to solid-planting, crop rotations etc. Where adequate shelter for the adult parasites and predators is lacking, this must be provided. It may be necessary to provide artificial sources of adequate carbohydrates and proteins, e.g. honey dew, nectar, pollen and other extra floral secretions. Natural enemies need to be protected by judicious use of selective and non-persistant insecticides, as most natural enemies are susceptible to broad-spectrum chemicals. This approach is most readily and effectively manipulated and involves recognition of realistic thresholds of damage, careful monitoring of pest populations and the use of insecticides only when necessary to prevent intolerable losses. Knowledge of the biology, ecology, and behaviour of the insect pest often permits application of insecticides to very restricted areas that the insect pests utilize for feeding, breeding or hiding (Bottrell, 1979). It may also be possible to detect and utilize insecticide resistant parasites and predators but so far little progress has been made in this direction. However, Croft (1976) presented a summary of basic biological, ecological and toxicological characteristics of three phytoseiid mite predators which have developed resistance to certain insecticides commonly used in commercial fruit orchards in countries such as Australia, Canada, United States of America and Netherlands.

For a comprehensive review of biological control including its methodology, reference should be made to the treatise by Huffaker & Messenger (1976).

8.3 Evaluation of control agent effectiveness

The effect of natural enemies on their prey populations is often extremely difficult to assess (Kiritani & Dempster, 1973). This is especially true for predation as here the prey is often completely consumed whereas the pathogens and parasites can be relatively easily identified within the body of a host. Repeated observations of the decline of pest populations following the introduction of natural enemies may not be enough to evaluate the effectiveness of control agents.

Experimental methods of evaluation, used routinely, are mainly based on elimination of natural enemies by various mechanical or chemical means, and comparison of pest populations in their presence or absence. Kiritani & Dempster (1973) refer methods of natural enemy evaluation to one of two categories:

(*i*) indirect methods of study;
(*ii*) direct methods of study.

In the indirect methods of evaluation, correlation is sought between the numbers, or rate of increase, of the prey and the numbers of the natural enemy. The direct methods may involve:

(*a*) exclusion of natural enemies by various methods;
(*b*) measurement of the action of the enemy;
(*c*) observation of the natural enemy action directly in the field.

Mechanical exclusion methods involving the use of devices such as cages, cloth sleeves, or wire screens have often been used to protect one of the experimental units from parasites and predators while in the other a natural enemy is allowed to operate freely. The difference in the host population usually gives a measure of natural enemy effect.

In chemical exclusion techniques insecticide is used in one experiment unit to eliminate or inhibit natural enemies while the other unit remains unsprayed. Again, the difference in host populations gives a measure of natural enemy effectiveness. This method has been successfully used by Owusu-Manu (1976) to evaluate the effectiveness of natural enemies in controlling the populations of *Bathycoelia thalassina*, a pest of cocoa in Ghana.

A biological check method utilizes the symbiotic association between some ants and homopterous scale insects. The latter feed the ants on their honeydew secretions while in return receiving protection from the attack of predators. Exclusion of ants through a barrier exposes the pest to natural enemy attack. A decline in pest population demonstrates the effectiveness of natural enemies. Major direct evaluation methods of natural enemies are compared in Table 6. As noted by Kiritani & Dempster (1973), no single technique is adequate for all situations and we are still a long way off reliable techniques for studying all enemy/prey interactions.

8.4 The economics of biological control

Simmonds (1968), in a detailed analysis of the economics of biological control, has shown that biological control has proved to be a highly profitable method in terms of costs and economic returns. According to him a total expenditure of £1 000 000 in seven successful biological control projects on various crops over a period of 40 years has yielded total benefits of £4 780 000. DeBach (1964) estimates that biological control projects at the University of California, with an aggregate investment of £1 500 000 during 1923–59, are currently providing an annual benefit of about £10 500 000 per annum. From examples given by authors such as Clausen (1956), DeBach (1964), Huffaker (1974, 1975) and Simmonds & Bennett (1977) etc., it is clear that biological control programmes have in certain instances completely eliminated the need for chemical applications. In making the cost benefit analysis, DeBach (1972) estimates that for each dollar spent on parasite introduction, there has been a benefit of $30 to agriculturists. On the other hand, each dollar spent on insecticides has returned only $5 in benefits (Pimental *et al.*, 1965). It is obvious that research into biological control has produced very favourable returns.

However, it must be admitted that biological control using parasites and predators has failed to control effectively many of the well known pests such as the European corn borer, gypsy moth, alfalfa weevil, etc. However, the use of biological control agents has now greatly reduced the magnitude of the gypsy moth problem while good control of alfalfa weevil has recently been achieved in the corn belt region. Kamran (1973) notes the dismal record of the introduction of neotropical tachinids into South East Asia for biological control of stem borers of graminaceous crops. He records that in the Philippines and India, despite years of painstaking work and numerous releases, the Cuban fly, *Lixophaga diatraeae* (Tachinidae), failed to survive on the borers. The introduction of parasites from India to control sugarcane stem borers in Ghana has been equally disappointing. Similarly, there does not appear to be a single case of successful control of rice stem borers using introduced parasites alone. Probably, as noted by DeBach (1964), from the successes obtained so far, it appears that biological control is strongly related to island ecology. However, the 'island theory'

Table 6 The comparison of major evaluation methods of natural enemies (after Kiritani & Dempster, 1973).

Method	Procedure for quantitative evaluation	Major limitations	Kinds of information obtained
Exclusion techniques			
Mechanical exclusion	Exclusion of natural enemies (n.e.) → Changes of prey number	(1) Change of microclimate (2) Interference of dispersal (3) Evaluation of individual n.e. is difficult	(1) Demonstrate the degree of control by n.e. complex as a whole (2) Gives no measure of density dependency nor information on compensatory factor
Insecticidal check and handpicking	Ditto	(1) Imperfect exclusion of n.e. (2) Faunal change by insecticide (3) Evaluation of individual n.e. is difficult in insecticidal check (4) Time-consuming for handpicking	The same as for mechanical check, except that hand-picking gives measure of compensatory factors when used appropriately.
Feeding trace method			
Precipitin test	Sampling of predator → % of positive predators → Detectable time of a meal; population size of predators	(1) Reduction of predators (2) Specificity of antiserum (3) Not appropriate to gregarious prey and voracious predator	Demonstrate predator species of the prey and relative importance of individual predator species (2) Quantitative evaluation of individual predator species is possible (3) Possible to examine the density-dependency of predation (4) Gives no information on compensatory factor.

Table 6 (*continued*)

Radionuclide labelling	Release of tagged prey → Sampling of predator → % of positive predators; amount of individual burden → Biological half-life; population size of predators; population size of prey	(1) Release of tagged prey (2) Emigration of tagged prey (3) Difficulty of interpretation of radionuclide burden	Theoretically the same for the precipitin test
Direct observation			
Sight-count	Frequency of predation → Diurnal rhythm of predation; time taken for consuming a prey	(1) Predation has to be observed (2) Likely disturbance by observation (3) Difficulty in obtaining a valid parameter of food retention time	The same for the precipitin test and demonstrate the predator species of the named prey species

put forth by Imms (1931) and Taylor (1955) which predicts that host suppression by natural control should be more successful on islands than on non-islands has been rejected by a number of workers notably Clausen (1958) who cites many successes on the North America mainland. According to Hall *et al*. (1980), this theory is inconsistent with empirical facts and should be rejected.

The significance of the type of host plant crop to successful biological control of insect pests has been discussed by Lloyd (1960). DeBach (1964) notes that biological control has generally been successful where the target insect is a pest of a perennial rather than annual crops but the latter provide most of the food for mankind. However, recently Sailer (personal communication) reports excellent biological control of the leaf beetle – a pest of an annual crop growing over a large continental area of the United States. Although control of many pests by biological means has not been successful, the inherent advantages to control by this method in terms of environmental and other considerations have led to considerable expansion of research in recent years. Still not enough is being done. Funding agencies are usually reluctant to support projects where quick returns are not forthcoming. Painstaking and detailed research, identifying and manipulating differences among different strains of natural enemies may very well, in some cases, mean the difference between success and failure in biological control.

8.5 Literature cited

Akanbi, M. O. (1978). Current status of biological control in Africa with special reference to forest pests in Nigeria. *PANS*, **24(2)**, 121–8.

Akotoye, N. A. K. & Kumar, R. (1977). Natural enemies of *Characoma stictigrapta* Hmps. (Lepidoptera: Noctuidae), a pod and leaf feeding pest of cocoa in Ghana. *Proc. 5th International Cocoa Res. Conf. Ibadan*, 1975, 443–7.

Allen, G. E., Gregory, B. G. & Brazzel, J. R. (1966). Integration of the *Heliothis* nuclear polyhedrosis virus into a biological control programme on cotton. *J. Econ. Ent.*, **59(6)**, 1333–6.

Andres, L. A. & Bennett, F. D. (1975). Biological control of aquatic weeds. *Ann. Rev. Ent.*, **20**, 31–46.

Andres, L. A. & McMurty, J. A. (1979). Introduction of new species and biotypes, pp. 41–5. In *Biological Control and Insect Pest Management.* Eds D. W. Davis, S. C. Hoyt, J. A. McMurty and M. T. AliNiazee. Division of Agricultural Sciences. University of California, Richmond, California. 102 pp.

Angus, T. A. (1968). The use of *Bacillus thuringiensis* as a microbial insecticide. *World Rev. Pest Control*, **7(1)**, 11–26.

Angus, T. A. (1977). Microbial control of Arthropod pests. *Proc. XV International Congr. Ent. Washington*, 473–7.

Annecke, D. P. & Moran, V. C. (1978). Critical reviews of biological pest control in South Africa. *J. ent. Soc. Afr.*, **41(2)**, 161–88.

Askew, R. R. (1971). Parasitic insects. American Elsevier Publications, New York. 316 pp.

Bolton, B. (1973). A bacterial pathogen of cocoa capsids. *Annual Report (1970-71). Cocoa Research Institute, Tafo, Ghana*. p. 155–6.

Bottrell, D. G. (1979). *Integrated Pest Management*. Council on Environmental Quality. U.S. Government Printing Office, Washington, D.C. 120 pp.

Bucher, G. E. (1964). The regulation and control of insects by fungi. *Ann. ent. soc. Queb.*, **9**, 30-42.

Burges, H. D. & Hussey, N. W. (eds) (1971). Microbial control of insects and mites. Academic Press, New York. 862 pp.

Burns, E. C., Wilson, B. H. & Tower, B. A. (1961). Effect of feeding *Bacillus thuringiensis* to caged layers for fly control. *J. Econ. Ent.*, **54(5)**, 913–15.

Charpentier, L. J., Jackson, R. D. & McCormick, W. J. (1973). Sugarcane borer control by delta endotoxin of *Bacillus thuringiensis* HD–1, in field tests. *J. Econ. Ent.*, **66(1)**, 249–51.

Clausen, C. P. (1956). Biological control of insect pests in the continental United States. *U.S. Dept. Agric. Tech. Bull.*, **113**, 1–151.

Clausen, C. P. (1958). Biological control of insect pests. *Ann. Rev. Ent.*, **3**, 291–310.

Creighton, C. S. & McFadden, T. L. (1974). Complementary actions of low rates of *Bacillus thuringiensis* and chlordimeform hydrochloride for control of caterpillars. *J. Econ. Ent.*, **67**, 102–4.

Compere, H. (1961). The red scale and its natural enemies. *Hilgardia*, **31**, 175–278.

Croft, B. A. (1976). Establishing insecticide-resistant phytoseiid mite predators in deciduous tree fruit orchards. *Entomophaga*, **21(4)**, 383–99.

David, W. A. L. (1975). The status of viruses pathogenic for insects and mites. *Ann. Rev. Ent.*, **20**, 97–117.

DeBach, P. (ed.) (1964). *Biological Control of Insect Pests and Weeds*. Chapman and Hall, London. 844 pp.

DeBach, P. (1972). The use of imported natural enemies in insect pest management ecology. *Proc. Tall Timbers Conf. Ecol. Animal Control Habitat Management*, **3**, 211–33.

DeBach, P. (1974). *Biological Control by Natural Enemies*. Cambridge University Press, Cambridge. 323 pp.

de Groot, P., Cunningham, J. C. & McPhee, J. R. (1979). Control of red-headed pine sawfly with a baculovirus in Ontario in 1978 and a survey of areas treated in previous years. *Canadian Forestry Service, Department of Environment, P.O. Box 2190, Sault Ste. Marie, Ontario, Canada*. Report FPM-X-20, 1–14.

Daoust, R. A. & Roome, R. E. (1974). Bioassay of a nuclear-polyhedrosis virus and *Bacillus thuringiensis* against the American bollworm, *Heliothis armigera* in Botswana. *J. invertebr Pathol.*, **23**, 318–24.

Dulmage, H. T. & Rhodes, R. A. (1971). Production of pathogens in artificial media, 507–39. In *Microbial Control of Insects and Mites.* Eds H. D. Burgess and N. W. Hussey. Academic Press, Inc., New York. 862 pp.

Dysart, R. J. (1973). The use of Trichogramma in the USSR. *Proc. Tall Timbers Conf. Ecol. Animal Control Habitat Management*, **4**, 165–73.

Evans, H. C. (1974). Natural control of arthropods, with special reference to ants (Formicidae), by fungi in the tropical high forest of Ghana. *J. appl. Ecol.*, **11**, 37–49.

Falcon, F. L. A. (1976). Problems associated with the use of arthropod viruses in pest control. *Ann Rev. Ent.*, **21**, 305–24.

Firempong, S. & Kumar R. (1975). Natural enemies of *Toxoptera aurantii* (Boy.) (Homoptera:Aphididae) on cocoa in Ghana. *Biol. J. Linn. Soc.*, **7(4)**, 261–92.

Fisher, R. & Rosmer, L. (1959). Toxicology of the microbial insecticide, Thuricide. *Agr. Food. Chem.*, **7**, 686–8.

Greathead, D. J. (1971). A review of biological control in the Ethiopian regions. *Technical Communication, Commonwealth Institute of Biological Control*, No. **5**, 1–162.

Hagen, K. S. & Bishop, G. W. (1979). Use of supplemental food and behavioral chemicals to increase the effectiveness of natural enemies, pp. 49–60. In *Biological Control and Insect Pest Management*. Eds D. W. Davis, S. C. Hoyt, J. A. McMurty and M. T. AliNiazee, Division of Agricultural Sciences, University of California, Richmond, California. 102 pp.

Hall, J. B., Kumar, R. & Enti, A. A. (1972. The obnoxious weed *Eupatorium odoratum* (Compositae) in Ghana. *Ghana Jnl. Agr. Sci.*, **5**, 75–8.

Hall, R. W., Ehler, L. E. & Bisabri-Ershadi, B. (1980). Rate of success in classical biological control of arthropods. *Bull. Ent. Soc. Amer.*, **26(2)**, 111–14.

Hassell, M. P. (1978). *The Dynamics of Arthropod Predator-prey System*.

No. 13 of monographs in population biology. Princeton University Press, Princeton, New Jersey. 198 pp.

Heimpel, A. M. (1967). A critical review of *Bacillus thuringiensis* var. *thuringiensis* Berliner and other crystaliferous bacteria. *Ann. Rev. Ent.*, **12**, 287–322.

Heimpel, A. M. (1977). The use of viruses in plant protection. *Pontificiae Academiae Scientiarum Scripta Varia*, **41**, 275–91. (Published as: Marini-Bettòlo, G. B. (ed.) (1977) Natural products and the protection of plants. Proceedings of a study week at the Pontifical Academy of Sciences. October 18–23, 1976. Elsevier Scientific Publishing Company, Amsterdam. 846 pp.)

Heinrich, B. & Bartholomew, G. A. (1979). The ecology of the African dung beetle. *Scientific American*, November 1979, 118–26.

Hoffman, R. A. & Gingrich, R. E. (1968). Dust containing *Bacillus thuringiensis* for control of chicken body, shaft and wing lice. *J. Econ. Ent.*, **61(1)**, 85–8.

Huffaker, C. B. (1959). Biological control of weeds with insects. *Ann. Rev. Ent.*, **4**, 251–76.

Huffaker, C. B. (1974). *Biological Control*. Plenum, New York and London. 511 pp.

Huffaker, C. B. (1975). Biological control in the management of pests. *Agro-Ecosystems,* **2**, 15–31.

Huffaker, C. B. & Messenger, P. S. (eds) (1976). *Theory and Practice of Biological Control.* Academic Press, New York. 788 pp.

Imms, A. D. (1931). *Recent Advances in Entomology*. Blakistons, Philadelphia. 374 pp.

Irabagon, T. A. & Brooks, W. M. (1974). Interaction of *Campoletis sonorensis* and a nuclear polyhedrosis virus in larvae of *Heliothis virescens. J. Econ. Ent.*, **67(2)**, 229–31.

Kamran, M. A. (1973). Introduction of neotropical tachinids into Southeast Asia for biological control of stem borers of graminaceous crops. *Bull. Ent. Soc. Amer.*, **19(3)**, 143–6.

Kiritani, K. & Dempster, J. P. (1973). Different approaches to the quantitative evaluation of natural enemies. *J. appl. Ecol*, **10(1)**, 323–30.

Kumar, R. (1979). Biological control of cocoa pests in West Africa. *Proc. 5th Conf. West African Cocoa Entomologists, Yaounde*, 1976, 109–16.

Kumar, R. (1981). The case for the establishment of an insect identification service and taxonomic research centre(s) in Africa. *Insect Sci. Application,* **1(4)**, 425–30.

Kumar, R. & Nutsugah, D. (1979). Parasite complex on two lepidopterous cocoa pests in Ghana. *Niger. Jour. Agric. Sci.*, **1(1)**, 45–9.

Lack, D. (1954). *The Natural Regulation of Animal numbers*. Clarendon Press, Oxford, 343 pp.

Lavabre, E. M., Ban, J. & Vandamme, P. (1966). Étude préliminaire de la transmission de viroses nucléaires et de bactérioses aux chenilles des cacaoyers. *Revue Café, Cacao Thé*., **10(4)**, 336–41.

Lewis, F. B., Dubois, N. R., Grimble, D., Meterhouse, W. & Quinby, J. (1974). Gypsy moth: efficacy of aerially-applied *Bacillus thuringiensis. J. Econ. Ent*., **67(3)**, 351–4.

Lewis, W. J., Jones, R. L., Gross, H. R., Jr. & Nordlund, D. A. (1976). The role of kairomones and other behavioral chemicals in host finding by parastic insects. *Behavioral Biol*., **16**, 267–89.

Lewis, W. J., Jones, R. L. & Sparks, A. N. (1972). A host-seeking stimulant for the egg parasite *Trichogramma evanescens*: its source and a demonstration of its laboratory activity. *J. Econ. Ent*., **65**, 1087–9.

Le Pelley, R. H. (1968). *Pests of Coffee*. Longmans, Green & Co., London. 590 pp.

Lloyd, D. C. (1960). Significance of the type of host plant crop in successful biological control of insect pests. *Nature*, **187(4735)**, 930–1.

Madelin, M. F. (1960). Internal fungal parasites of insects. *Endeavour*, **19**, 181–90.

Mistric, M. J. & Smith, F. D. (1973). Tobacco budworm: control on flue-cured tobacco with certain microbial pesticides. *J. Econ. Ent*., **66**, 979–82.

Moran, V. C. (1981). Belated kudos for cochineal insects. *Antenna*, **5(2)**, 54–7.

Moran, V. C. & Annecke, D. P. (1979). Critical reviews of biological pest control in South Africa. 3. The jointed cactus, *Opuntia aurantica* Lindley. *J. ent. Soc. S. Afri.*, **42(2)**, 299–329.

Morris, O. N. (1977). Long term study of the effectiveness of aerial application of *Bacillus thuringiensis* – acephate combination against the spruce budworm, *Choristoneura fumiferana* (Lepidoptera : Tortricidae). *Canad. Entom.*, **109**, 1239–48.

Nguyen-Ban, J. (1975). Note préliminaire sur l'élevage en laboratoire de Trichogrammes en prévision d'une lutte intégrée contre *Earias biplaga* (Walk.) (Lepidoptera : Noctuidae), *Proc. 4th Conf. West African Cocoa Entomologists, Legon*, 160–4.

Norris, J. R. (1968). Biological methods of insect control. *PANS*, **14**, 505–13.

Owusu-Manu, E. (1976). Natural enemies of *Bathycoelia thalassina* (Herrich-Schaeffer) (Hempiptera: Pentatomidae), a pest of cocoa in Ghana. *Biol. J. Linn. Soc.*, **8(3)**, 217–44.

Pimental, D., Chant, D. A., Kelman, A., Metcalf, R. L., Newsom, L. D. & Smith, C. N. (1965). Improved pest control practices, p. 227–91. In *Restoring the Quality of our Environment.* Report of the Environmental Pollution Panel of the President's Science Advisory Committee. The White House. Washington, D.C. 317 pp.

Poinar, G. O., Jr. & Thomas, G. M. (1978). *Diagnostic Manual for the Identification of Insect Pathogens*. Plenum Press, New York. 232 pp.

Ridgway, R. L. & Vinson, S. B. (1977). Biological control by augmentation of natural enemies. Insect and mite control with parasites and predators. *Proc. Symp. XV International Congr. Ent. Washington*. Environmental Science Research, Volume 11. Plenum Press, New York. 480 pp.

Rockwood, L. P. (1950). Entomogenous fungi of the family Entomophthoraceae in the Pacific Northwest. *J. Econ. Ent.*, **43**, 704–7.

Roome, R. E. (1975). Field trials with a nuclear polyhedrosis virus and *Bacillus thuringiensis* against larvae of *Heliothis armigera* (Hb.) (Lepidoptera, Noctuidae) on sorghum and cotton in Botswana. *Bull. ent. Res.*, **65**, 507–14.

Sailer, R. I. (1975). Parasite release programs – pros and cons. *Proc. FAO/IAEA Training Course on use of Radioisotopes and Radiation in Entomology*, 193–204.

Sailer, R. I. (1976). Future role of biological control in management of pests. *Proc. Tall Timbers Conf. Ecol. Animal Control Habitat Management*, **6**, 195–209.

Sailer, R. I. (1979). Use of biological control agents. pp. 139–151. In *Introduction to Crop Protection.* Ed. W. B. Ennis, Jr. American Society of Agronomy and the Crop Science Society of America, 677 South Segoe Rd., Madison, U.S.A.

Schröder, D. (1970). Memorandum on the possibilities of biological control of some important insect pests and noxious weeds in West African Commonwealth countries (especially in Ghana). *Rep. Commonw. Inst. Biol. Control Kumasi* (cyclostyled). 72 pp.

Sheldrick, R. D. (1968). The control of Siam Weed (*Eupatorium odoratum* Linn.) in Nigeria. *J. Niger. Inst. Oil Palm Res.*, **5**, 7–19.

Shervis, L. J. & Shenefelt, R. D. (1973). A controlled indexing vocabulary for *Apanteles* species literature. *Bull. Ent. Soc. Amer.*, **19(3)**, 147–52.

Simons, W. R. (1978). Preliminary research on entomophagous nematodes in particular on *Neoaplectana* species in the Netherlands. *Med. Fac. Landbouww. Rijksuniv. Gent.*, **432**, 756–8.

Simmonds, F. J. (1967). The economics of biological control. *J. Royal Soc. Arts.*, Oct. 1967, 880–98.

Simmonds, F. J. (1968). Economics of biological control. *PANS*, **14(A)**, 207–215.

Simmonds. F. J. (1972). Biological control in the tropics. *Technical Bulletin, Commonwealth Institute of Biological Control*, No. 15, 159–169.

Simmonds, F. J. & Bennett, F. D. (1977). Biological control of Agricultural pests. *Proc. XV International Congr. Ent. Washington*, 464–72.

Somerville, H. J. (1977). The insecticidal endotoxin of *Bacillus thuringiensis. Pontificiae Academiae Scientiarum Scripta Varia*, **41**, 253–68.

Steinhaus, E. A. (1949). *Principles of Insect Pathology*. McGraw-Hill Book Co, Inc., New York, N.Y. 757 pp.

Steinhaus, E. A. (1967). Microbial control – a comment on its present status in the United States. *Bull. Ent. Soc. Amer.*, **13(2)**, 104–8.

Stinner, R. E. (1977). Efficacy of inundative releases. *Ann. Rev. Ent.*, **22**, 515–31.

Summers, M., Engler, R., Falcon, L. A. & Vail, P. (1975). *Baculoviruses for Insect Pest Control: safety considerations.* American Society of Microbiology, Washington, D. C. 186 pp.

Sutter, G. R. (1969). Treatment of corn rootworms larvae and adults with *Bacillus thuringiensis* and *B. popilliae. J. econ. Ent.*, **62(3)**, 756–7.

Tanda, Y. (1959). Microbial control of insect pests. *Ann. Rev. Ent.*, **4**, 277–301.

Taylor, T. A. (1974). Evaluation of Dipel for control of Lepidopterous pests of Okra. *J. econ. Ent.*, **67(5)**, 690–1.

Taylor, T. H. C. (1955). Biological control of insect pests. *Ann. appl. Biol.*, **42**, 190–6.

Tinsley, T. W. (1979). The potential of insect pathogenic viruses as pesticidal agents. *Ann. Rev. Ent.*, **24**, 63–87.

Tooke, F. G. C. (1955). The eucalyptus snout-beetle, *Gonipterus scutellatus* Gyll. A study of its ecology and control by biological means. *Ent. Mem. Dept. Agric. S. Afr.*, No. 3, 282 pp.

Waterhouse, D. F. & Wilson, F. (1968). Biological control of pests and weeds. *Science Journal*, **4(12)**, 31–7.

Wilson, F. (1964). The biological control of weeds. *Ann. Rev. Ent.*, **9**, 225–44.

Wilson, B. H. & Burns, E. C. (1968). Introduction of resistance to *Bacillus thuringiensis* in a laboratory strain of house flies. *J. econ. Ent.*, **61(6)**, 1747–8.

9 Genetic Control

This form of control, a relatively recent development in economic entomology, involves the use of genetically impaired pests to limit reproduction and survival of their own species in natural populations. Such pests are mass produced in the laboratory and released among wild populations in the field so that mating with normal insects will either not result in an offspring or lead to reduced fitness (e.g. sterility, failure to adapt properly to the environment etc.) of the progeny.

9.1 Methods of genetic control

Autocidal approaches have received considerable attention during the last 20 years, they have been discussed at length by Davidson (1974) and several reviews on different aspects of the subject are available (for example, see Proverbs, 1969; Pal & LaChance, 1974; North, 1975 and Whitten & Foster, 1975). These methods may be discussed under the following headings:

(1) Sterile-insect release method
(2) Chromosomal translocations
(3) Hybrid sterility
(4) Cytoplasmic incompatibility

9.1.1 Sterile-insect release method (SIRM)

The basis of this method, also termed sterile-male technique, lies in the fact that male insects can be easily sterilized and when released mate with wild fertile females whose sterile eggs contribute nothing to, and thereby effectively reduce, the population of the progeny. In practice, while rearing and sterilizing insects on a large scale, it has seldom been practicable to distinguish male insects from females and thus both have been treated and sterilized and released in the field. However, recent developments have included the use of lasers to distinguish the sexes and machines to separate them. Thus it may be possible, at least in some species, to release only males. If the process of release of sterile insects is repeated at suitable intervals, it should theoretically be possible to eradicate the pest population (Knipling, 1955, 1959). For success with this method, the following two principal conditions must be met:

(1) It must be possible to mass-rear the target species at an acceptable cost;
(2) The insect produced should be sterile yet competitive, i.e. the procedure employed for sterilizing the males must not adversely affect the mating behaviour of the insects. It is to be noted that procedures will probably affect some or all aspects of mating but the method used must produce an acceptable insect.

Sterilization by radiation is known to produce in some insect species, traits such as lessened flight capability, insufficient mobility, etc. (see Offori, 1970 for references). However, with carefully controlled doses of radiation it is now possible to overcome such problems in at least some species (see Hooper, 1976 for use of nitrogen to lessen the deleterious effects of sterilizing doses of gamma radiation on male competitiveness in some insects). At low dosage it may be possible to induce a dominant lethal mutation in the sperm that would still be competitive or have an acceptable level of competence. Hooper & Katiyar (1971) in comparing the results of the interaction between sexual competitiveness and sterility of males of the Mediterranean fruit-fly (*Ceratitis capitata*), irradiated with various doses of gamma radiation, found that as the dose increased male competitiveness decreased. It is desirable that the female mates only once but this does not appear to be a critical criterion for the success of the method. According to Knipling (1960), monogamous mating by the female is not a requirement, provided the sterile sperms are fully competitive with normal sperm when in the female genital tract. The level of the population of the pest to be controlled should be low so that the release of a proportionately large number of irradiated males among the population would be economically feasible. This may be achieved better either by releases when the pest population is normally low or the population may be lowered by some other means of control, e.g. by the use of insecticides. There must be a cheap and rapid method for the efficient dispersion of the sterilized insects. This may involve the use of aircraft to distribute large numbers of sterile insects uniformly and rapidly over large areas of difficult terrain. The degree of competitiveness of irradiated males is of crucial importance in eradication or suppression programmes (Knipling, 1955; Waterhouse *et al.*, 1976). If, as a result of sterilization, the competitiveness of the insect is reduced, then the number of insects must be increased to attain the overflooding ratio necessary to produce a reduction in the target population. Larger releases may not always solve the problem as there is evidence from some Lepidoptera that sterile males do not elicit normal reproductive behaviour in their mates and are not therefore competitive (see Cutkomp, 1967 for requirements for success with radiation induced sterility).

Radiation-sterilization for the eradication of pests was conceived by E. F. Knipling in the U.S.A. in 1937. The classic example of the use of sterilized insects is the successful eradication from the island of Curacao of the screw-worm, *Cochliomyia hominivorax*, a fly with larvae parasitic on cattle (Baumhover *et al.*, 1955). The programme required the production of 170 000 screw-worm flies per week with a release of up to 800 sterile flies per 2.59 km^2 per week. This caused a marked decline in the wild populations and eradication was complete within 6 months. Since then screw-worm has been eradicated in Florida, and controlled in south-west United States. But migration from Mexico into the border states has necessitated the creation of a 'sterility barrier' zone to a width of 483 km. In this zone some 200 million sterile flies are released weekly. Despite set-backs during 1972–76, attributed to loss of reproductive compatibility between the sterilized strain and the wild flies, releases of sterile *C. hominivorax* have continued in the barrier zone, stopping fertile flies from re-populating the U.S.A. Cost-benefit ratio of the programme has been estimated to range from \$39 to \$113 benefit for each dollar spent depending on how many factors are taken into consideration (see LaChance, 1979 for further information on this aspect). A programme to eradicate screw-worm from Mexico is currently in progress.

SIRM has also been successfully used with other Diptera, e.g. the melon fly, *Dacus cucurbitae*, was eradicated from the island of Rota, near Hawaii, by the release of 257 million flies irradiated as pupae with a cobalt source (Steiner *et al.*, 1965). This fly as well as *D. dorsalis* were also eradicated from Guam by SIRM. Mediterranean fruit-fly (*Ceratitis capitata*) has been eradicated from the United States six times by means of insecticides, but re-invasion occurred near Los Angeles in 1975. To avoid further use of insecticides, it was decided to adopt SIRM against the 1975 outbreaks. By releasing over 600 million sterile medflies in the Los Angeles area between October 1975 and May 1976, the outbreak was eradicated saving citrus crops worth over $1.6 billion per year (LaChance, 1979).

Since 1966, in the Tijuana area of California, 3/4 million radiation-sterilized Mexican fruit-flies (*Anastrepha ludens*) have been released per week to protect the soft fruit industry. Indeed, the release of these insects has now replaced the use of insecticides. Since 1975, an additional 3/4 million sterile fruit-flies per week have been released in southern Baja for the same purpose. Altogether, over 600 km^2 of prime fruit-producing country are protected for 7 months each year at an annual cost of $266 000 (LaChance, 1979).

The integrated use of insecticides, sex pheromone traps, biological control with phytoseiid mites, and SIRM against the codling moth (*Cydia pomenella*) has been suggested for trials in North America (LaChance, 1979). In Africa, considerable research is currently underway to control and eliminate the populations of tsetse flies, *Glossina* spp., by SIRM. Spectacular results should however not be expected until a lot more research work has been done on the problem.

SIRM has the capacity of becoming increasingly efficient with successive applications. It has what is termed the 'bonus effect'. Table 7 shows estimated effects of different control measures on a hypothetical insect population.

Many workers have attempted to apply the SIRM concept to depress populations of other insects without much success. Two groups of insects, Hemiptera and Lepidoptera, which include some of the major pests in the world, are relatively more radio-sensitive than other insects. The inability of irradiated male Lepidoptera to compete successfully with native males for females has been suggested as the principal reason for the lack of success (North & Holt, 1971). Further, most Lepidoptera are polygamous which complicates competition. Recently Toba *et al.* (1972) have obtained better control of the populations of caged

Table 7 Relative trends of hypothetical populations subject to different systems of control (after Knipling, 1966, slightly modified).

Generation	Uncontrolled population[1]	Insecticide treatments (90% kill each generation)	Chemosterilant treatments (90% sterility each generation)	Sterile insect release (9 000 000 each generation)	Integrated programme of insecticides and and sterile insect releases
Parent	1 000 000	1 000 000	1 000 000	1 000 000	1 000 000
F_1	5 000 000	500 000	50 000	500 000	45 450
F_2	25 000 000	250 000	25 000	131 625	9880
			125	9540	485
F_3	125 000 000	125 000	6	50	0
F_4	625 000 000	62 500	0	0	
F_5	3 125 000 000	31 250			
F_6	15 625 000 000	15 625			
F_7	78 125 000 000	7812			
F_8	390 625 000 000 (eventually limited by density factors)	3906			
	Total requirements for theoretical elimination of populations =	Treatments for 20 generations	Treatments for 5 generations	45 000 000 sterile insects	Insecticide treatments for 1 generation plus 4 000 000 sterile insects

[1] It is assumed that the uncontrolled population increases at a fivefold rate until the maximum density for the environment is reached, which is assumed to be 125 000 000.

cabbage loopers by the release of substerilized insects. The differences have been attributed to the fact that F_1 progeny exhibit a higher degree of sterility than was induced in their parents. This approach obviously significantly lowers cost of the control measures. In another experiment when adult pink bollworms irradiated with a substerilizing dose (10 krad) of gamma irradiation (Bariola *et al*., 1973) were released in field cages and in a 5-acre field of cotton, larvae with chromosomal aberrations that could cause sterility in the adult stage were found in both cages and the field. When 3500 partially sterile moths (mixed sexes) were released in a 1/200-acre cage over a 7-week period, sufficient larvae did not develop to cause significant damage to the cotton.

The use of gamma radiation to sterilize insects has many draw-backs in addition to those stated earlier, chief amongst which are:

(1) highly skilled scientists are required to operate the programmes;
(2) investment costs are higher in terms of equipment and facilities for radiation treatments than for conventional control measures. It must however be emphasized that as long as there is a favourable cost/benefit ratio, e.g screw-worm project (see above) high cost *per se* is not a problem.
(3) biology and behaviour of the insects may be adversely affected by radiation treatments.

These difficulties have led to search for chemical sterilants for use in the laboratory and field and the characteristics of these have been discussed by Bŏrkovec (1966) whilst Campion (1971) has provided a useful summary of attempts to use these agents in pest control. The affects of certain chemosterilants on the adult males of the flour moth (*Ephestia kuehniella*) have recently been reported by Tan & Mordue (1977). Chemosterilants have been used with success though not on any substantial scale in the field, against house-flies (*Musca domestica*), screw-worm flies, and various fruit-flies (Tephritidae) (La Brecque, 1968). By weekly spraying of garbage dumps in Italy with 0.0625–0.2% tepa in 1% malt bait it has also been found possible to suppress house-fly populations for as long as treatments are maintained. Because this approach is open to the objection that mutagenic aklylating agents are being disseminated, hempa was also evaluated in the same way and proved successful at concentrations of 1.25–3.75% in the bait. Likewise on an island in Japan, hempa baits were employed for effective experimental control of *M. domestica* in 1967. It is difficult to see why there has been little subsequent work with this approach to the problems of pest control.

Techniques for producing sterile insects Various types of ionizing radiations have been used to sterilize insects. These include alpha, beta and gamma radiations of radioactive substances, X-rays and neutrons. Gamma and X-rays with their great penetrating powers are most useful for inducing sterility. Sterilization in insect control is effected by gamma rays usually ^{60}Co but occasionally ^{137}Ce. X-ray machines overheat and burn out during prolonged operations and are in addition expensive and cannot treat the large number of pupae at a time which is required in large programmes.

The other practical method of inducing insect sterility is by chemical treatment. Many chemosterilants are now available, they are usually classified as follows:

(1) Alkylating agents which are highly reactive and induce permanent sterility (tepa, metapa, thiotepa, melphalan).
(2) Antimetabolites, which are structurally related to biologically active substances, are most effective against females and have temporary action (purine, pyrimidine analogues).
(3) Miscellaneous compounds such as herbicides, hormones, antibiotics and fungicides.

It must however be remembered that most of the chemosterilants are highly toxic to mammals and must be used with caution. They also affect the treated insects adversely so doses have to be adjusted in such a way that the desired degree of sterility is obtained without affecting the vitality of the released insects. In most species males and females react differently to the treatment. A lot of research is therefore necessary for each target insect proposed for genetic control.

9.1.2 *Chromosomal translocations or genetic load*

The normal arrangement of chromosomes can be broken by radiation or chemical treatment. The abnormal rearrangement or translocation does not normally cause sterility, though the fecundity of the insects may be reduced to varying degree, but the offspring carry the translocations and are likely to be sterile when outcrossed. Translocation strains have reduced fertility, the degree of sterility being related to the number of translocations. Nevertheless they can be bred in the laboratory with difficulty. Offspring of these strains when crossed with normal strain are fully sterile. Whitten (1970) has suggested that certain genes (e.g. insecticidal susceptibility) could be transported into natural populations by the release of laboratory-reared insects if these outnumber the natural population and are of equivalent viability. It is known that such synthetic strains are self propagative and may: *i*) render the populations amenable to conventional form of control (see Davidson, 1974, and Curtis, 1979, etc., *ii*) remove the noxious features of the pest such as vector capacity (see Davidson, 1974, and Curtis, 1979, etc.), *iii*) introduce deleterious traits, e.g. production of an excess of males (see, for example, Suguna *et al*., 1977) impaired longevity, production of a proportion of non-viable eggs, or *iv*) replace the natural population with one of reduced viability and therefore lower the population density.

In order to be able to utilize genetic methods such as translocations, a good knowledge of the genetics of the species concerned is essential (this is however, not a requirement for SIRM and hybrid sterility). Formal genetics is known extensively only in certain groups of insects. Thus two-thirds of Pal and Whitten's (1974) book on the use of genetics in insect control deals with dipteran species whose genetics is well known and for which genetical control methods have been quickly developed. On the other hand a great deal of work remains to be done on the other orders of insects. Detailed information on the ecology, population dynamics, mating, field behaviour of the species to be controlled is also necessary for this purpose (Bottrell, 1979).

Use of chromosomal rearrangements to alter the genetic composition of a mature population of a particular pest is reviewed by Foster *et al*. (1972). Progress in the use of chromosomal translocations for the control of insect pests has been documented by Robinson (1976). Curtis (1979) has also briefly reviewed the subject. It is now believed by some geneticists that a drastic reduction in the population of any insect may have unfortunate consequences on the ecosystem and that it would be desirable to introduce traits such as lack of vector capacity so that while still being in our environment, they cease to be a problem.

9.1.3 Hybrid sterility

Some strains of a given species show mutual incompatibility, and the eggs of such matings are infertile; the introduction of alien strains thus promote suppression of a population. Davidson (1969) stated that crossing of races, subspecies or sibling species often resulted in production of hybrid progeny that are sterile but vigorous. The sterile hybrids can be reared and released to eradicate the natural pest populations.

Laster *et al*., (1978) produced *Heliothis* hybrids from crosses between *Heliothis subflexa* females and *H. virescens* males. Male progeny from these crosses were sterile and females were fertile. When the hybrid females were backcrossed to *H. virescens* males, the sterile male trait persisted through successive backcross (BC) generations. *H. virescens* moths were released in field cages with BC_3 and BC_6 moths in 1:1 and 1:5 ratios respectively, to study infusion of the hybrid sterile-male trait into the *H. virescens* population. Egg hatch was reduced through sterile male matings and the hybrid sterile-male trait infused into the *H. virescens* populations. If this infusion holds true in natural populations, the sterile hybrid release concept should be an effective method of controlling the tobacco budworm, *H. virescens*.

Instances of satisfactory success in the control of insect pests by hybrid sterility are as yet lacking in the literature. However, Davidson (1974) believes that in the notorious African malaria and filariasis vector, *Anopheles gambiae*, which is a complex of some six sibling species, there is the potential of employing the sterile hybrid males as a measure of genetic control of natural populations. Crosses between any of the species of the complex produce sterile hybrid males but in some crosses there is also a shift in the sex ratio in favour of males. One particular cross between two of the species, species B (said to be *An. arabiensis*) males and *An. melas* females, has resulted in an abnormally high ratio of males to females (Davidson, 1974). A field experiment using these sterile males against a target population of *An. gambiae* was carried out at Pala, Upper Volta. A total of nearly 296 000 hybrid pupae (93% of these were males) were liberated into natural breeding sites of *An. gambiae*, the only member of the complex present in the area. The released sterile males dispersed quite well in the village environment, both indoors and outdoors but failed to mate with the wild females. The reasons for this failure were subsequently attributed to the presence of a behavioural pre-mating barrier. The disappointing results of this call for more basic research on the *An. gambiae* complex, before a second field trial is carried out (Davidson, 1974). Perhaps the target species should be represented in the parents of the hybrid males, preferably as the male parent.

9.1.4 Cytoplasmic incompatibility

Davidson (1974) defines cytoplasmic incompatibility as the situation where 'a cross between two populations of apparently the same species results in insemination without fertilization though with partial embryonation in some ova'. The spermatozoa enter the cytoplasm of the egg, but no actual fusion between spermatozoan nucleus and ovum takes place. Pal & LaChance (1974) state that a cytoplasmic factor, possibly rickettsiae living in the cytoplasm, inactivates the sperm of the incompatible male after entry into the egg. This phenomenon has been best studied in mosquitoes of the *Culex pipiens* complex where a number of crosses produce none or very few offspring which are invariably female, and of the same crossing type as their mother. This suggests the possibility of releasing an incompatible strain to eliminate wild populations provided the release material belongs to one sex only (males in case of mosquitoes as they neither bite nor transmit disease and are polygamous). Field trials with the use of this method were carried out in a village near Rangoon in Burma and several locations in India. In the latter country, through the use of a suitable crossing scheme, it was possible to introduce the entire chromosomal complement of a Delhi strain of *Culex pipiens fatigans* into Paris cytoplasm (Krishnamurthy & Laven, 1976).

The resulting stock was said to have competed fairly well for mating after release in the Delhi area (Grover *et al.*, 1976). The results of the release though indicating a degree of suppression of natural population, have not been followed further. Some work has been done on the *Aedes scutellaris* complex, the members of which are vectors of filariasis. Attempts to cross some of them have demonstrated that mating barriers between them are not complete. Females of *Ae. s. scutellaris* from New Guinea, crossed with *Ae. s. katherinensis* males from northern Australia, produced fertile offspring, while the reciprocal produced no progeny at all. F_1 generation of the crosses behaved like its maternal parent with the males remaining incompatible with female *Ae. s. katherinensis* (Woodhill, 1950). Davidson (1974) believes that mass-reared *Ae. s. scutellaris* males could be used to eradicate *Ae. s. katherinensis*. Published evidence suggests that the phenomenon of cytoplasmic incompatibility is quite complicated and calls for considerable research.

9.2 Future of genetic control

Despite the difficulties and drawbacks mentioned with the use of genetic control approaches, there is no reason to rule out genetical methods as impractical and academic. Indeed some very large programmes are currently in progress. These include the eradication of melon fly from southern islands of Japan; eradication of screw-worm from Mexico, and containment of Mediterranean fruit-fly in Mexico. The genetic methods have the following main attractions (Cutkomp, 1967):

(1) The pest control measures are conducted over large areas under professional supervision thereby avoiding mistakes or omissions by individual growers.

(2) Genetic control methods are potentially of increasing efficiency as the densities of target populations decline. Insecticidal control measures usually become less cost-effective as the target population decreases and there is an economic threshold, below which it becomes uneconomic to employ insecticides at all.

(3) Genetic control is specific and avoids undesirable effects on other organisms. No residues are involved and other adverse effects associated with the use of insecticides are avoided.

(4) Once established, it may cause dramatic economic savings by completely eliminating the pest.

Used together with other control agents in integrated pest management schemes, genetic control methods have a hopeful future. In a recent review of the use of SIRM for suppression or eradication of fruit-fly populations in Australia, Hooper, (1978) has expressed optimism in the integration of this method with other ecologically acceptable control approaches. Recently, in one such programme (Van der Vloedt *et al*., 1980), 16 km of the fringing forest of upper River Guénako in Upper Volta were treated with the synthetic pyrethroid decamethrin applied from a helicopter as an aerosol at a dosage of 0.2 g ha^1. Two applications at 14-days interval reduced the target *Glossina palpalis gambiensis* population by over 95%. Subsequently over-flooding of the residual *G.p.gambiensis* population with sterile males was easily achieved. Thus for vectors, SIRM offers the hope of species eradication. For agricultural pests, this method provides the proven ability, especially when integrated with other control methods, to suppress the target population well below the economic threshold level, at reasonable operational costs. The use of autocidal approaches in pest control is likely to increase in the future.

9.3 Literature cited

Bariola, L. A., Bartlett, A. C., Staten, R. T., Rosander, R. W. & Keller, J. C. (1973). Partially sterilized adult pink bollworms: releases in cages and field cause chromosomal aberrations. *Environmental Entomology*, **2(2)**, 173–6.

Baumhover, A. H., Grahm, A. J., Bitter, B. A., Hopkins, D. H., New, W. D., Dudley, F. H. & Bushland, R. C. (1955). Screw-worm control through release of sterilized flies. *J. Econ. Ent*., **48**, 462–6.

Bǒrkovec, A. B. (1966). Insect chemosterilants. *Advances in Pest Control. Research.* vol. VII. Interscience Publishers. 143 pp.

Bottrell, D. R. (1979). *Integrated Pest Management.* Council on Environmental Quality: U.S. Government Printing Office, Washington, D.C. 120 pp.

Campion, D. G. (1971). Chemosterilisation. *PANS*, **17(3)**, 308–14.

Curtis, C. F. (1979). Translocations, hybrid sterility, and the introduction into pest populations of genes favourable to man. pp. 19–30. In *Genetics in Relation to Insect Management.* Eds M. A. Hoy and J. J. McKelvey Jr. The Rockfeller Foundation, 1133 Avenue of the Americas, New York. 179 pp.

Cutkomp, L. K. (1967). Progress in insect control by irradiation induced sterility. *PANS*, **13(1)**, 61–70.

Davidson, G. (1969). The potential use of sterile hybrid males for the eradication of member species of the *Anopheles gambiae* complex. *Bull. World Health Organ.*, **40**, 211–8.

Davidson, G. (1974). *Genetic Control of Insect Pests.* Academic Press, London. 158 pp.

Foster, G. G., Whitten, M. J., Prout, T. & Gill, R. (1972). Chromosome rearrangements for the control of insect pests. *Science*, **176(4037)**, 875–80.

Grover, K. K., Curtis, C. F., Sharma, V. P., Singh, K. R. P., Dietz, K., Agarwal, H. V., Razdan, R. K. & Vaidyanathan, V. (1976). Competitiveness of chemosterilized males and cytoplasmically incompatible translocated males of *Culex pipiens fatigans* Wied. *Bull. ent. Res.*, **66**, 469–80.

Hooper, G. H. S. (1976). Sterilization of *Dacus cucumis* French (Diptera: Tephritidae) by gamma radiation. III. Effect of radiation in nitrogen on sterility, competitiveness and mating propensity. *J. Aust. Ent. Soc.*, **15**, 13–18.

Hooper, G. H. S. (1978). The sterile insect release method for suppression or eradication of fruit fly populations. In *Economic Fruit Flies of the South Pacific Region.* Watson Ferguson & Co., Brisbane, Australia. 137 pp.

Hooper, G. H. S. & Katiyar, K. P. (1971). Competitiveness of gamma-sterilized males of the Mediterranean fruit fly. *J. Econ. Ent.*, **64(5)**, 1068–71.

Knipling, H. F. (1955). Possibilities of insect control or eradication through the use of sexually sterile males *J. Econ. Ent.*, **48(4)**, 459–62.

Knipling, H. F. (1959). Sterile male method of population control. *Science*, **130**, 902–4.

Knipling, H. F. (1960). The eradication of the screw-worm fly. *Sci. Amer.*, **203(4)**, 54–61.

Knipling, H. F. (1966). Plant protection in the American economy. The entomologist's arsenal. *Bull. Ent. Soc. Amer.*, **12(1)**, 45–51.

Krishnamurthy, B. S. & Laven, H. (1976). Development of cytoplasmically incompatible and integrated (translocated incompatible) strains of *Culex pipiens fatigans* for use in genetic control. *J. genet.*, **62**, 117–29.

LaBrecque, G. C. (1968). Laboratory procedures, pp. 31–98. In *Principles of Insect Chemosterilization*. Eds G. C. LaBrecque and C. N. Smith. Appleton Century Crofts, New York. 354 pp.

LaChance, L. E. (1979). Genetic strategies affecting the success and economy of the sterile insect release method, pp. 8–18. In *Genetics in Relation to Insect Management*. Eds M. A. Hoy and J. J. McKelvey, Jr. The Rockfeller Foundation, 1133 Avenue of the Americas, New York. 179 pp.

Laster, M. L., Martin, D. F., Pair, S. D. & Furr, R. E. (1978). Infusion of hybrid *Heliothis* male sterility into *H. virescens* populations in field cages. *Environmental Entomology*, **7(3)**, 364–6.

North, D. T. (1975). Inherited sterility in *Lepidoptera. Ann. Rev. Ent.*, **20**, 167–82.

North, D. T. & Holt, G. G. (1971). Inherited sterility and its use in population suppression of Lepidoptera, pp. 99–111. In *Application of Induced Sterility for Control of Lepidopterous Populations. Int. Atomic Energy Agency Publications STI/PUB/281, Vienna, Austria.* 169 pp.

Offori, E. D. (1970). Gamma irradiation of *Stomoxys calcitrans. J. Econ. Ent.*, **63(2)**, 574–9.

Pal, R. & LaChance, L. H. (1974). The operational feasibility of genetic methods for control of insects of medical and veterinary importance. *Ann. Rev. Ent.*, **19**, 269–91.

Pal, R. & Whitten, M. J. (Eds) (1974). *The Use of Genetics in Insect Control.* American Elsevier Publ. Co., Inc., N.Y. 241 pp.

Proverbs, M. D. (1969). Induced sterilization and control of insects. *Ann. Rev. Ent.*, **14**, 81–102.

Robinson, A. S. (1976). Progress in the use of chromosomal translocations for the control of insect pests. *Biol. Rev.*, **51**, 1–24.

Steiner, L. F., Harris, E. J., Mitchell, W. C., Fujimoto, M. S. & Christenson, L. D. (1965). Melon fly eradication by overflooding with sterile flies. *J. Econ. Ent.*, **58(3)**, 519–22.

Suguna, S. G., Curtis, C. F., Kazmis, S. J., Singh, K. R. P., Razdan, R. K. & Sharma, V. P. (1977). Distorter double translocation heterozygote systems in *Aedes aegypti. Genetica*, **47**, 117–23.

Tan, K. H. & Mordue, W. (1977). Effects of certain chemosterilants and biologically active substances on the adult Mediterranean flour moth, *Ephestia kuehniella* Zeller (Lepidoptera: Pyrallidae). *Bull. ent. Res.*, **67**, 483–9.

Toba, H. H., Kishaba, A. N. & North, D. T. (1972). Reduction of populations of caged cabbage loopers by release of irradiated males. *J. Econ. Ent.*, **65(2)**, 408–11.

Van der Vloedt, A. M. V., Baldry, D. A. T., Politzar, H., Kulzer, H. & Cuisance, D. (1980). Experimental helicopter applications of decamethrin followed by release of sterile males for the control of riverine vectors of trypanosomiasis in Upper Volta. *Insect Sci. Applications*, **1**, 105–12.

Waterhouse, D. F., LaChance, L. E. & Whitten, M. J. (1976). Use of autocidal methods, pp. 637–59. In *Theory and Practice of Biological Control.* Eds C. B. Huffaker and P. S. Messenger. Academic Press, New York. 788 pp.

Whitten, M. J. (1970). Genetics of pests in their management. pp. 119–35. In *Concepts of Pest Management.* Eds R. L. Rabb & F. E. Guthrie. North Carolina University, Raleigh, U.S.A. 242 pp.

Whitten, M. J. & Foster, G. G. (1975). Genetic methods of pest control. *Ann. Rev. Ent.*, **20**, 461–76.

Woodhill, A. R. (1950). Further notes on experimental crossing within the *Aedes scutellaris* group of species (Diptera: Culicidae). *Proc. Linn. Soc. N.S.W.*, **75**, 251–3.

10 Chemical Control

10.1 Meaning of insecticide

Literally, insecticide means 'insect killer'. Although the name seems to suggest that only insects are killed, in reality this is not so. Insecticides are chemicals that affect the biological processes of many living organisms and may thus act as poisons to many animal species. It is this property of insecticides that is utilized in killing insects and other pests. Broadly speaking, insecticides are chemicals used to combat pests.

10.2 Uses of insecticides

Insecticides are at the moment man's chief weapon against insect pests. The large-scale use of agricultural chemicals is already one of the main factors in minimizing losses due to pests. Insecticides have been used most heavily in U.S. agriculture on a few major crops. The percentage of total insecticides used on them is as follows:
cotton 47%, corn 17%, fruits 9%, vegetables 7%, soybeans 4%, and tobacco 3% (Metcalf, 1980). Gross returns on U.S. investment in chemical control are estimated at $8.7 billion or more (Pimentel *et al.*, 1978). Braunholtz (1979) estimates that the world crop protection market has grown from £500 million (M) grower prices in 1960, 1750 M in 1970 to £4200 M in 1978. During this period, insecticide sales are reported to have grown by 13% to £1500 M; herbicides by 12% to £1700 M; fungicides by 8% to £750 M, and plant growth regulators etc. by 11% to £250 M (Braunholtz, 1979). Most experts agree that a removal of insecticides from crop protection would result in an immediate drop in food supplies (NAS, 1975). Discontinuation of all pesticide usage, it has been estimated (Braunholtz, 1979), would reduce the production of crops and livestock by 30% and would increase the price of farm produce by 50–70%. In Britain,

similar action would reduce the cereal crops by 45%, potato crop by 42% and sugar beet by 67% (Braunholtz, 1979). According to Anonymous (1972), in 1968 the use of pesticides on cotton and cocoa in West African countries produced extra crops to the value of $139 million, i.e. 5–10% of the exports of the region as a whole. At least five African nations are dependent on cotton for a large percentage of their export earnings (Cunningham, 1973) and insecticides are a vital factor in cotton growing. However, artificial or real shortages, soaring prices, lack of appreciation of pest problems by several governments are having a serious retarding effect on insecticide use and adversely affecting crop production in many countries in Africa. While grave concern has been expressed over the side effects of the use of chemicals, 90% of insecticide usage throughout the world is effective and achieves the object (of its use) quickly and efficiently. Although the search for alternative methods of pest control continues at an accelerated pace, it is difficult to think of a time when it will not be necessary to make some use of insecticides for the control of pests.

10.3 Development and marketing

Before the Second World War, the chemicals used for destroying pests were largely inorganic chemicals such as compounds of lead and arsenic which are well known poisons. Some organic chemicals of plant origin, e.g. nicotine, pyrethrum and rotenone were known as well, but their use was limited because of high production costs. The era of synthetic and cheap chemicals began during the Second World War with the discovery of DDT. Paul Muller received a Nobel Prize for medicine in 1948 for his synthesis of DDT. A big spur to synthetic insecticides was also given by the discovery that some products of the diene reaction with chlorinated intermediates were active insecticides. This led to the synthesis of aldrin and dieldrin which, until recently, were some of the major insecticides in use in the world.

The development of an insecticide is a costly operation. One of the first stages is the screening for biological activity of a vast number of chemicals, generally the products of major industries, such as the petroleum industry. The promising chemicals are further tested on rats and other vertebrates to determine their toxicity (see Whitney, 1969; Anonymous,

1971*a*; Braunholtz, 1977, 1979; Corbet, 1979, etc. for an insight into the development of modern agro-chemicals). Published statistics show that in 1977 it was necessary to screen over 12 000 compounds to get a pesticide on the market, whereas in 1970 the figure was 7400, and in 1956 only 1800 (Gilbert, 1978). It has been estimated that the research and development of a new insecticide from primary screening to significant sales in a major market, takes at least 7–8 years, and now costs about ten million pounds sterling (Braunholtz, 1979). This excludes investment in manufacturing plant and due to recent world-wide inflationary trends, soaring energy prices, the costs are steadily rising. Even so, by the time the insecticide reaches the market, it may have become obsolete due to the discovery of better chemicals elsewhere. In view of the amount of money required and the research involved, often using sophisticated equipment, and the uncertainty of returns, progress in the development of better chemicals has not been as fast as it could have been. Because of our continued dependence on insecticides, it is now being suggested that research into the development of new materials which are more specific, more toxic to target insects and less toxic to other organisms, particularly man and his associated animals, will in the near future have to be supported, at least partially, by government or related agencies.

10.4 Ideal insecticide

The ideal insecticide is a chemical which stays at the place of application through its active period; is toxic to particular pests but harmless to other organisms including man; is easily used; is able to break down into harmless products in the environment within a reasonable time, and must at the same time be cheap to produce. No chemical with all these properties is known at present.

10.5 Names of insecticides

An insecticide may have three different names:

(1) Trade or proprietary names under which the chemical is marketed. These are the most commonly-used names and several trade names may be assigned to one chemical.

(2) A name which denotes the chemical structure of the insecticide.
(3) A common name approved by a national or international organization, e.g. International Organization for Standardization. Sometimes codes are also used. These usually consist of a company initial plus serial number.

10.6 Major groups of insecticides

Several books dealing with insecticides are available (see for example, Brown, 1951; Metcalf, 1955; Gunther & Jeppson, 1960; O'Brien, 1967; Brooks, 1974; Matsumura, 1975; Metcalf & McKelvey, 1976, etc.). The *Pesticide Manual* (Worthing, 1979) gives invaluable information on nomenclature, history, manufacture, the physical, chemical and biological properties, formulations and methods of analysis of these compounds. An index of chemical, common and trade names of pesticides has been compiled by Mercer (1981). A handbook useful to workers in West Africa is by Adeyemi (1970). Although providing information on agricultural insecticides available in Nigeria, it is generally applicable to other parts of West Africa and supplies information on insecticide properties, toxicity, suitability, dose rates, safety and first aid measures. Techniques for testing insecticides have been reviewed by Busvine (1971*a*).

The principal groups of insecticides currently in use in Agriculture and Public Health are:

(1) Chlorinated hydrocarbons, e.g. DDT, Chlordane, BHC, Aldrin, Dieldrin, Endrin and Toxaphene (Figs 9 and 10).
(2) Organophosphates, e.g. Parathion, TEPP, Malathion, and Diazinon (Figs 11 and 12).
(3) Carbamates, e.g. Sevin, Ortho-Bux, Elocron and Baygon (Fig. 13).
(4) Organic insecticides of plant origin, e.g. Pyrethrum, Rotenone, etc (Figs 14 and 15).
(5) Certain dinitro and fluorine compounds, and a variety of special fumigants such as ethylene dibromide and dichloride, are also used as insecticides.
(6) Hormones and pheromones are also considered as insecticides, but these are dealt with separately.

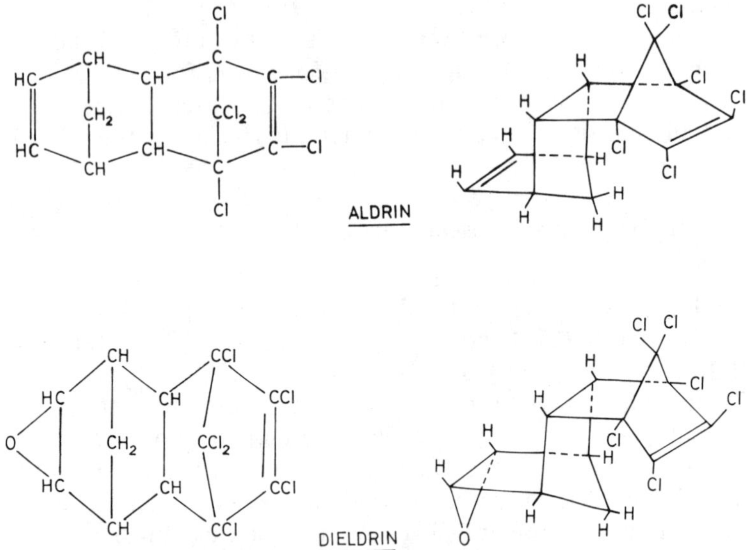

ALDRIN

DIELDRIN

Figure 9 Aldrin; dieldrin.

10.6.1 Chlorinated hydrocarbons or organochlorine compounds (often abbreviated as O.C.)

This was probably the most widely used group of insecticides. They are fairly complex, stable compounds of low volatility, and their insecticidal activity is believed to depend on the arrangement of chlorine atoms in the molecule. They are among the more persistent insecticides, i.e. they leave residues, and so their insecticidal effects last longer. Most have a moderate mammalian toxicity. Traces accumulate in the bodies of many vertebrates, including man, and this has caused concern over their long-term effects. These compounds have many properties in common, and insects which become resistant to one member of the group generally, but not always, become resistant to all, i.e. a strain develops which is generally no longer killed by any of the insecticides of this group. The following three chemical classes of organochlorines have been extensively used in crop protection. Of all the compounds available for pest control, chlorinated hydrocarbons such as DDT are perhaps the cheapest

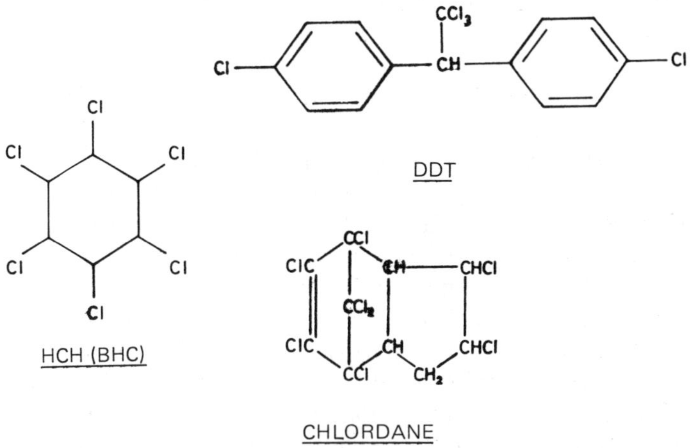

DDT

HCH (BHC)

CHLORDANE

Figure 10 Chlordane; Benzene hexachloride; DDT.

and at the same time effective killers and therefore have been widely used for destroying insect pests. Their use has now diminished greatly but they are still in use in several countries in Africa.

Cyclodienes These insecticides are nearly all manufactured by Diels-Alder condensation of a cyclic diene with a dienophile. They include aldrin, dieldrin, endrin and endosulfan. These are used in public health and crop protection, especially against soil-inhabiting pests and, depending upon the environmental conditions, can persist in the soil for several years. In West Africa, aldrin and dieldrin are mainly used against termites and the variegated grasshopper.

As with other chlorinated hydrocarbons, traces of cyclodienes have been detected in the bodies of many vertebrates and their use in the United Kingdom and several other countries has been restricted since the mid-sixties.

Bridged diphenyls Examples are DDT, Dicofol, Rhothane and Chlorbenside. Many analogues of DDT, e.g. methoxychlor, have been synthesized. DDT has good residual effect and when first introduced in 1940, it was used mainly against vectors of disease. Since then it found a multiplicity of uses in agriculture where it has been mainly used against foliage feeding insects, particularly those injurious to fruit trees. In view of its high persistence, a single application is often adequate to control many pests. However, to reduce hazards from environmental contamination, including its storage in mammalian tissues, use of DDT in agriculture and horticulture in several European countries and in the United States of America has been illegal since at least 1970. In Canada, it has been replaced by other groups of insecticides. In West Africa it is used against pests of Kenaf and a variety of caterpillars, including stem borers. World Health Organization (WHO) still recommends the continued use of DDT for indoor spraying to control medical vectors while restricting all outdoor uses, especially application to water (Tahori & Galun, 1976).

Benzene hexachloride (*BHC*) This compound is also known as HCH. Its isomer, Lindane, is a very useful general purpose insecticide and was first introduced around 1945. There is some evidence to indicate that under tropical conditions, because of its relatively high vapour pressure, its residual action is not as effective as under temperate conditions. It has been extensively used in West Africa against cocoa mirids and some other pests but due to the development of resistance, it is being replaced by other insecticides. BHC is also known to accumulate in mammalian tissues and thus its use against pests of cattle and on pastures has been discontinued since the early 1960s. For a detailed treatment of chlorinated hydrocarbons see Brooks (1974).

10.6.2 Organophosphorous compounds (*often abbreviated as O.P.*)

These are esters or organic salts of phosphoric acid or its derivatives. Some are derived from compounds which were developed during the Second World War in the search for potent warfare agents. Generally these compounds are highly toxic to mammals as well as the target organisms although some less toxic to mammals are known. As a rule these insecticides are

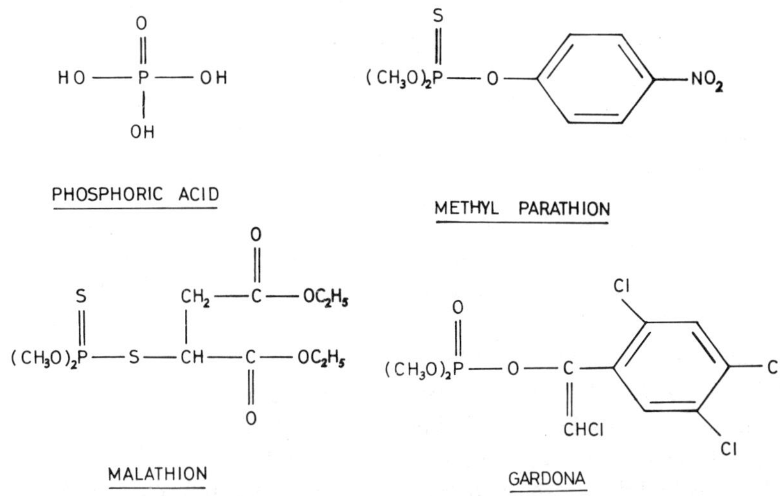

PHOSPHORIC ACID

METHYL PARATHION

MALATHION

GARDONA

Figure 11 Phosphoric acid; Methyl parathion; Malathion; Gardona.

much less stable than the organochlorines, their persistence varying from about a day to several months. Many of these compounds, e.g. bidrin, possess systemic properties while others are fumigants. The systemic insecticides, after absorption into plants, are translocated to various parts where they function as a stomach poison, killing pests that feed on them. These insecticides may persist for several weeks inside the plants before being degraded by hydrolysis or other plant enzyme action. Such insecticides are highly effective against sap sucking and chewing insects and usually do not harm non-target organisms such as parasites, predators and honey-bees. Organophosphorous compounds are currently quite widely used and well known insecticides included in this group are: Malathion, Parathion, Bromophos, Monocrotophos, Methidathion and Tetrachlorvinphos. Essential features and uses of organophosphorous insecticides are discussed by Braunholtz (1968) and Eto (1974). The use of systemic insecticides in warm climates has been discussed by Leatherdale (1966).

10.6.3 Carbamates (often abbreviated as Carb.)

In the last 20 years, a great deal of research has been conducted

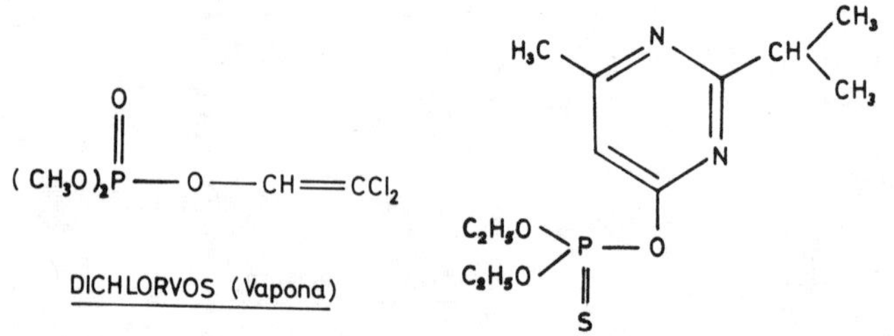

DICHLORVOS (Vapona)

DIAZINON (0.0 – diethyl-0-(2-isopropyl-6-methyl-4-pyrimidinyl) -thiophosphate)

Figure 12 Dichlorvos (Vapona); Diazinon.

on the N-methyl-carbamate insecticides. They are based on carbamic acid NH_2 COOH. The introduction and successful development of carbaryl led to the development of the insecticidal potential of other carbamates. In general these compounds are stable although in the environment in formulations they easily break down. Further they are inexpensive and relatively broad spectrum insecticides. They have a rapid yet transient poisoning effect in mammals: Weiden & Moorefield (1965) used the term 'selective spectrum' with the carbamates to denote their safety to mammals and other beneficial organisms. The carbamate insecticides are currently being used widely in insect control. This usage has undoubtedly resulted in part from the development of resistance to other types of insecticides such as the organophosphates.

Among the well known carbamate insecticides mention may be made of Carbaryl (Sevin) and Propoxur (Baygon). Since 1960 Sevin has been successfully used commercially on a number of pests of cotton, deciduous and small fruits, vegetable crops, cereals and several other crops as well as household pests. Baygon was introduced about mid-sixties as a highly effective agent against household pests such as cockroaches. Since then it has been found useful against a large number of crop pests and very recently it has proved highly successful against cocoa-mirids.

The history and development of carbamates have been reviewed by Davies (1968) while the relationships between the structure and the activity of a large number of synthetic carbamates are analysed by Metcalf (1971). A recent book on

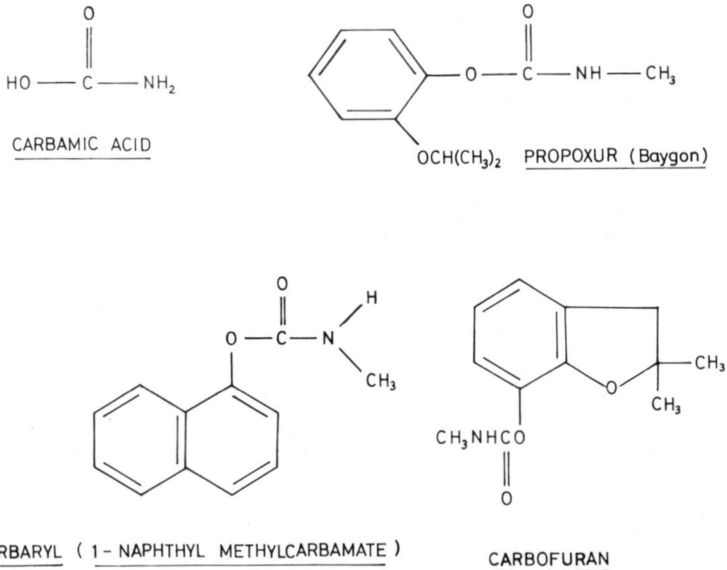

CARBAMIC ACID

PROPOXUR (Baygon)

CARBARYL (1-NAPHTHYL METHYLCARBAMATE)

CARBOFURAN

Figure 13 Carbamic acid; Propoxur (Baygon); Carbaryl; Carbofuran.

the chemistry, biochemistry and toxicology of carbamate insecticides is by Kuhr & Dorough (1976).

10.6.4 Pyrethroids

The use of pyrethrum powder, an extract from the flower heads of *Chrysanthemum cinerariaefolium*, is of considerable antiquity (Casida, 1973). Kenya, Tanzania, and Ecuador cultivate large areas of pyrethrum for commercial extraction of the natural insecticide. This is a mixture of six insecticidal esters: Pyrethrin I, II, Cinerin I, II, Jasmolin, I, II. The structures of these natural products, elucidated some 20 years ago, has recently been fully confirmed in every stereochemical detail by modern spectroscopic methods (see Elliott & Janes, 1979 for references). The term 'pyrethrins' is now the collective trivial name accepted internationally for the insecticidal compounds present in the flowers of pyrethrum. The natural pyrethrins have attracted attention because they are powerful insecticides but possess a very low mammalian toxicity. They have been used for many years mainly for the control of public health and veterinary vectors. The Produce Inspection Division of the Ghana Cocoa Marketing Board routinely performs evening fogging of cocoa sheds with 0.5% natural pyrethrins to control insect pests, especially the tropical warehouse moth, *Ephestia cautella*. The fogging exercise has to be carried out every evening or at least every other day in view of the short-lived effect of the pyrethrins. Although the use of pyrethrins has depended on the rapid knock down of pests, their application has been restricted due to lack of prolonged residual action and because they are expensive. However, the present concern for contamination of the environment by the use of insecticides has resulted in search for alternative compounds and this includes some painstaking research on pyrethrum extracts.

Modern studies on extracts of *C. cinerariaefolium*, especially by Elliott and co-workers at Rothamsted, have provided a considerable family of synthetic analogues, the modern pyrethroids, which constitute commercially important and useful materials. The initial aim of the synthetic studies was to

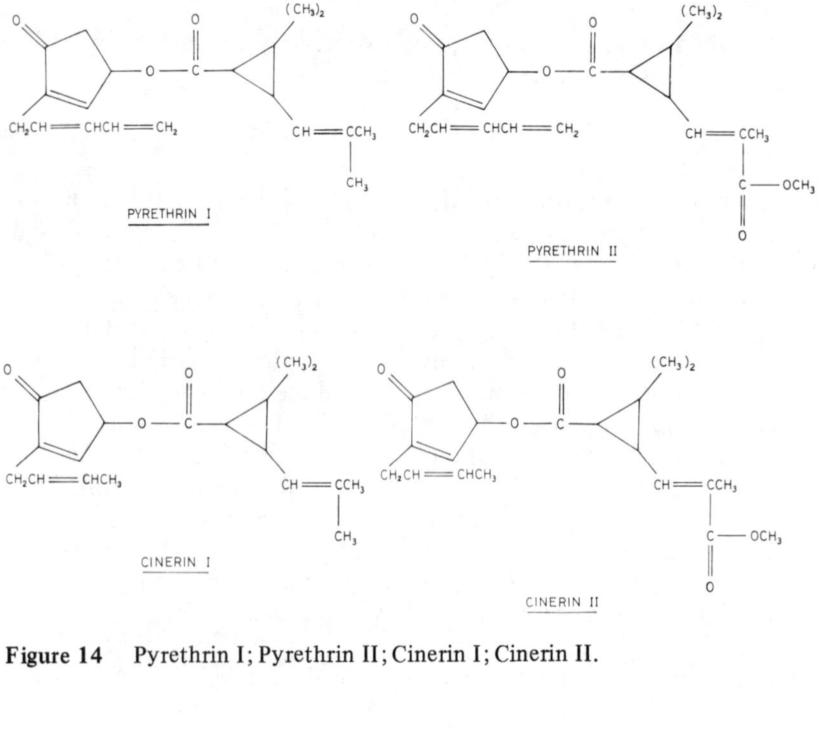

PYRETHRIN I

PYRETHRIN II

CINERIN I

CINERIN II

Figure 14 Pyrethrin I; Pyrethrin II; Cinerin I; Cinerin II.

produce non-persistent analogues which possessed low mammalian toxicity and were safe to use on food and edible crops. This was accomplished by Elliott *et al.* (1967) with the development of resmethrin and bioresmethrin which demonstrated that the synthetic products might prove more effective than the naturally-occurring insecticides. Attention was subsequently directed towards the development of compounds with greater insecticidal activity or faster knockdown than the natural esters, enhanced photostability and diminished mammalian toxicity. This research resulted in the development of permethrin (NRDC 143) (Elliott *et al.*, 1973) and decamethrin (NRDC 161) (Elliott *et al.*, 1974). Another interesting discovery is the highly active substance fenvalerate (of Sumitomo Chemical Company) which substitutes an isopropyl group for the dimethyl cyclopropane ring and a p-chlorophenyl group for the unsaturated cyclopropane constituent. Indeed the ester linkage is almost the sole resemblance to the original pyrethrin structure. Their virtual insolubility in water and their rapid and strong absorption to the soil organic matter tends to greatly reduce their mobility in the environment (Breese & Searle, 1977). In the past, as indicated above, the use of natural pyrethrins or their synthetic analogues in the field had been limited because of their rapid breakdown under field conditions. NRDC 143 is 10–100 times more stable in light than the previous pyrethroids. Widespread testing has confirmed NRDC 161 as the most powerful insecticide known for wide range of insect species (Elliott, 1979). The modern synthetic pyrethroids are said to show a level of stability which compares favourably with and often surpasses that of organophosphorous insecticides (Breese & Searle, 1977; Elliott, 1977*a*). For a progress report on synthetic pyrethroids reference should be made to the symposium volume on the subject edited by Elliott (1977*b*). Elliott & Janes (1979) state that some of the synthetic pyrethroids have more favourable properties than other groups such as the organochlorine compounds because, 'although they persist adequately on crop surfaces, their physical properties restrict migration in solution and as vapour, and they are rapidly decomposed when exposed to metabolizing systems, such as soil micro-organisms'. Investigations in the use of these chemicals against pests are currently in progress. Highly

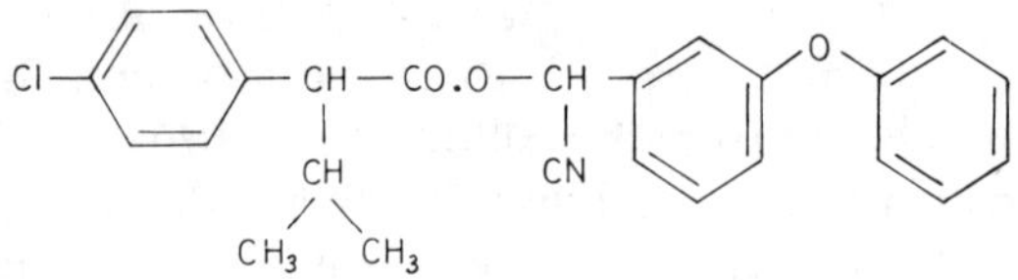

Fenvalerate (α – cyano – m – phenoxybenzyl
α – isopropyl – p – chlorophenylacetate)

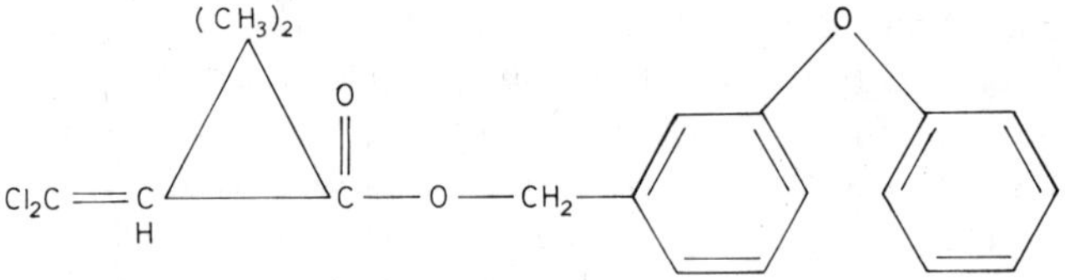

PERMETHRIN

Figure 15 Fenvalerate; Permethrin.

promising and commercial control, often superior to that provided by other widely-used insecticides has been obtained. Shell Chemicals recently launched Belmark (common name fenvalerate) as a cotton pesticide in Central and South America. It is said to provide good control of a number of traditional cotton pests. In some trials the number of formulations was halved (Anonymous, 1977). Highwood (1980) reports that results from a number of field trials on cotton in Malawi, Colombia and Egypt and sweet peppers in the U.S.A., using synthetic pyrethroids such as cypermethrin and fenvalerate all provide 'evidence to suggest that pyrethroid treated crops give yields higher than might be expected from examination of pest counts'. He has speculated that at commercially used dosages some pyrethroids might be less hazardous to parasites and predators. Increased yields could thus be the result of integration of chemical and biological control. Highwood also suggests that the increased yield might be due to a simulating effect of the pyrethroid on the crop.

10.6.5 Other botanical insecticides

Apart from pyrethrum, several naturally-occurring pesticides, in many cases biodegradable, are known. These include the following:

Nicotine This is an alkaloid prepared from waste tobacco, *Nicotiana tabacum* or *N. rustica*. This is a non-persistent, non-systemic, contact insecticide and an effective fumigating agent with some ovicidal properties.

Rotenone This insecticide forms component of the root of certain *Derris* and *Lonchocarpus* species. It is a selective, non-systemic insecticide with some acaricidal properties, and has low persistence but no phytotoxicity.

Ryania This is the insecticidal component of the ground stemwood of a shrub, *Ryania speciosa*. Ryanodine, the water soluble insecticidal component is a selective stomach insecticide that has been effectively used against lepidopteran larvae.

In addition to the above mentioned botanical pesticides, a number of plant products are known. For example, citronella oil is an insect repellent; pisatin from garden peas is a fungicide; extracts from some *Brassica* species are plant growth regulators. Actually over 2000 higher plant species have been shown to have some insecticidal properties. Elliott (1979) states that 'in many fields, not only that of insecticides, the most active synthetic compounds are often derived from the structure of chiral, natural products.' Further discovery and exploitation of such prototypes should obviously be rewarding. However, considerable research and financial investment is required in this field to characterize new and novel insecticides.

10.7 Mode of action of insecticides

Most modern insecticides act on the nervous system and enter the insect body either by contact or through the gut or by having a fumigant action (Table 8). Several insecticides enter the body in more than one way. Nervous activity involves the transmission of nerve impulses along nerve fibres (axons) and

Table 8 Some technical information on certain common insecticides.

Group	Insecticides	Contact	Stomach	Fumigant	Residual and action persistence	Mammalian toxicity	Systemic action
	Nicotine $C_{10}N_{14}N_2$ (Alkaloid from tobacco)	×	×	×	None	High	None
Plant origin (Botanical insecticides)	Pythrethrum (From flowers of *Chrysanthemum*)	×	—	—	None	Slight	None
	Rotenone $C_{23}H_{22}C_6$ (From roots of certain *Derris* and *Lonchocarpus* spp.)	×	—	—	Low persistence	Relatively harmless to most mammals	None
	Ryanodine $C_{25}H_{35}NO_9$ (Alkaloid from stemwood of *Ryania speciosa*)	×	—	—	Low	Slightly irritant to mammals	None
	Sabadilla (Alkaloid from the seeds of *Schoenocaulon officinale*)	×	—	—	None	None	None
Chlorinated hydrocarbons (Synthetic organic insecticides)	DDT $C_{14}H_9C_{15}$	×	×	—	High persistence	Moderate	None
	Aldrin $C_{12}H_8Cl_6$	×	×	Relatively volatile	Persistent	High	None

Table 8 *(continued)*

	BHC $C_6H_6C_6$	×	×	×	Persistent	Generally low and varies with isomers	None
	Chlordane $C_{10}H_6C_8$	×	×	×	High persistence	Low	None
	Dieldrin $C_{12}H_8Cl_6O$	×	×	—	Persistent	High	None
	Endrin $C_{12}H_8Cl_6O$	×	×	—	Persistent	High	None
	Toxaphene $C_{10}H_{10}Cl_8$	×	×	—	Persistent	Pronounced	None
Organophosphates (Synthetic organic insecticides)	Diazinon (Basudin) $C_{12}H_{21}N_2O_3PS$	×	—	×	—	Low	None
	Dichlorvos (Vapona) $C_4H_7Cl_2O_4P$	×	×	×	—	Low	None
	Dicrotophos (Bidrin) $C_8H_{16}NO_5P$	—	—	—	Moderate	High	Yes
	Fenitrothion $C_9H_{12}NO_5PS$	×	—	—	—	Low	None
	Malathion $C_{10}H_{19}O_6PS_2$	×	—	—	—	Moderate	None

Table 8 (*continued*)

Group	Insecticides	Contact	Stomach	Fumigant	Residual and action persistence	Mammalian toxicity	Systemic action
	Parathion $C_{10}H_{14}NO_5PS$	×	×	—	—	Very high	None
	TEPP (tetraethyl pyrophosphate) $C_8H_{20}O_7P_2$	×	—	—	—	Very high	None
Carbamates (Synthetic organic insecticides	Carbaryl (Sevin) $C_{12}H_{11}NO_2$	×	—	—	—	Low	Slightly systemic
	Carbofuran (Furadan) $C_{12}H_{15}NO_3$	—	—	—	—	Low	Yes
	Dioxacarb (Elocron) $C_{11}H_{13}NO_4$	×	—	—	—	Low	None
	Propoxur (Baygon, Unden) $C_{11}H_{15}NO_3$	×	—	—	—	Low, highly toxic to bees	None
Synthetic pyrethroids	Bioresmethrin $C_2H_{26}O_3$	×	—	—	Rapidly decomposed in air or light	None	None
	Fenvalerate $C_{25}H_{22}ClNO_3$	×	—	—	Photostable	Low	None

Table 8 *(continued)*

	NRDC 161 (Decamethrin*) $C_{22}H_{19}Br_2NO_3$	×	×	—	Photostable	Moderate	None
	Permethrin $C_{21}H_{20}Cl_2O_3$	×	—	—	Largely photostable	Low	None
	Resmethrin $C_{22}H_{26}O_3$	×	—	—	Rapidly decomposed in air and light	None	None
Fumigants of aliphatic chains	Ethylene dichloride $C_2H_4Cl_2$	—	—	×	—	High	None
	Ethylene dibromide $C_2H_4Br_2$	—	—	×	—	High	None
	Methyl bromide CH_3Br	—	—	×	—	High	None
	Naphthalene $C_{10}H_8$	—	—	×	—	Low	None

*According to Worthing (1979), the name decamethrin proved to be unacceptable and should not be used.

across the synapse to another nerve cell by means of a transmitter substance. The major transmitter substances are adrenalin and acetylcholine and these are normally inactivated by specific enzymes. Nerve poisons interfere with neuron or with synaptic transmission at either the delivery or target side of the synapse. However, the mode of action of most, if not all, insecticides is not fully understood and requires considerable fundamental work on the physiology of insects. One has only to examine the book edited by Wilkinson (1976) on insecticide biochemistry and physiology, in order to appreciate the complex problems involved in such studies. In both insects and mammals nerve impulses are carried across synapses and from nerve to muscle by a chemical, its transmitter. In normal animals, once a nerve impulse has been conducted across the gap by the transmitter substance say acetylcholine, the cholinesterase enzyme quickly acts to remove the acetylcholine. Normally such reactions happen very rapidly. However, organophosphorous compounds, on entering the body, become attached to the esteratic site of cholinesterase and prevent them from removing the acetylcholine. The accumulation of excessive amounts of this substance in the affected organs in insects results in rapid twitching of muscles, followed by paralysis and eventually death. In vertebrates two kinds of symptoms result: (*i*) twitching, and in severe cases convulsion; and (*ii*) lacrimation, salivation, narrow pupil, slow heart beat, and in severe cases, respiratory failure. In the case of acute poisoning, the central nervous system is also affected. Neurotoxicity is a dangerous complication caused by fluorinated phosphates or phosphanates. Neurotoxicity strikes several days after contact and results in irreversible destruction of axons and hence paralysis. Fortunately, commercial OP insecticides are not of these types. Antidotes are atropine, which blocks the synapses and thus counteracts the over-excitation; and oximes which hydrolyse the organophosphate–cholinestrase bond.

Carbamates They affect cholinesterase like OP but enter into binding with cholinesterase at both esteratic and anionic sites. This binding is however less stable, the poisoning is severe but of short duration. Atropine is antidotal as for OP poisoning.

However oximes are not capable of hydrolysing the carbamate–cholinesterase bond, hence they must not be given.

O'Brien (1978) stated that it is now almost universally agreed that the OPs and carbamates act by inhibiting the enzyme cholinesterase. According to him 'the older literature contains a number of protests against this apparently simplistic view', but none of the other explanations has been shown to be of importance in explaining the mode of action of these two groups of insecticides.

Organic insecticides of plant origin

Nicotine Causes synaptic blocking by competing with acetylcholine and leading to ganglionic stimulation followed by inhibition, resulting in central respiratory failure.

Pyrethrum Pyrethrins, as stated earlier, also act on nerve axons in DDT-like way. Elliott & Janes (1979) state 'although it is accepted that pyrethroids interfere with nerve action, the precise system attacked in insects is not known, neither is their mode of action well understood'. Earlier, O'Brien (1978) stated that pyrethroids appeared to vary in the relative importance of axonic and ganglionic effect in poisoned insects. Lesser toxic compounds such as barthrin may be largely axonic agents while fast knock-down compounds such as tetramethrin may have both ganglionic and axonic effects in affected insects. Pyrethrins in sublethal doses are metabolized, detoxified or excreted by insects. Low mammalian toxicity in pyrethroids has been attributed to the presence of sites in the molecule susceptible to metabolic attack, oxidation or hydrolysis. Pyrethrin (I) molecule is oxidized mainly at one site in the acid and at the diene side chain on the alcohol. According to Elliott (1979), bioresmethrin, which is even less toxic to mammals, has more susceptible sites in the alcohol, as well as being more vulnerable to esterases. Further, the ratio of toxicities to insects and mammals is said to indicate that the practical level of safety is more favourable for pyrethroids such as biopermethrin and bioresmethrin than for pyrethrin (I) (Elliott, 1979).

10.8 Consequences of the use of insecticides

We have witnessed over two decades of the most effective control of insect pests that man has ever known but this has brought a number of problems as discussed below. Some authors – Tahori (1971 *a*, *b*), Fletcher (1974), Moriarty (1975) and Perring & Mellanby (1978), have examined various facets of pest control and the effects of pesticides on the environment.

10.8.1 Need for re-application

Generally application of insecticides does not result in a permanent lowering of the numbers of a pest population to levels such that they no longer constitute an economic problem. Even after a careful use of insecticides, some of the pests in a population may not be killed and may remain in the field, multiplying again in the following season to pest proportions if insecticide applications are stopped. This requires annual recurring expenditure.

10.8.2 Effects on non-target organisms

Insecticides, as stated earlier, affect the biological processes of many living organisms and may thus act as poisons to a large number of animals, apart from the target species. In some places where chemicals were sprayed against insects, entire populations of birds were wiped out or significantly reduced, fish populations drastically lowered and residues of chemicals detected in human fat and the meat and milk of his cattle. Carson (1962), in her famous book *Silent Spring*, lucidly sets out details of the results of thoughtless and blanket spraying. Indeed environmentalists have frequently presented us with visions of the earth as a polluted planet where insecticides stand out as one of the chief culprits.

Effects of pesticides on non-target organisms, especially populations of parasites and predators, are now well recognized and documented (Ripper, 1956; Newsom, 1967; Coaker, 1977). In West Africa, the use of benzene hexachloride has led to the destruction of natural enemies of pests in the cocoa-ecosystem and caused once insignificant insect species to multiply to pest proportions (Owusu-Manu, 1976; Kumar, 1979). DeBach & Bartlett (1951) studied the effects of insecticides on the natural enemies of citrus and found that increase in the populations of each pest was correlated with a decrease in natural enemy numbers. Long term investigations in Nova Scotia, Canada, extending over a period of about 20 years, have indicated that 53 pest species of apple orchards, are kept under effective control by natural enemies. Extensive use of insecticides on cotton in the southern United States has led to the emergence of the red spider mite as an economic pest and these may be so numerous as to be found on 'practically every leaf'.

Indeed, wherever detailed studies have been made on the influence of insecticides on natural enemies of pests, the evidence is overwhelming that insecticides invariably upset the pest–natural enemy relationship, leading not only to the problems of primary pest resurgence and secondary pest outbreaks but also resulting in direct hazards to applicators and farm workers. This may be not only from mishandling and accidents but from exposure to contaminated surfaces where insecticide has been sprayed.

10.8.3 Problems of residues

Many insecticides, especially organochlorines such as DDT, leave residues in terrestrial and aquatic biota, cause cumulative food-chain concentration and biological magnification (see, for example, Hickey *et al.*, 1966; Lincer, *et al.*, 1981). Residues may have adverse effects on ecosystems by creating disequilibrium on food chains, scavengers, insect–host relationships, insect–plant relationships, etc. Further, a substantial part of applied chemical and its degraded products may persist for years in the bodies of animals, including man. Quite small concentrations may have substantial biological consequences. They may cause cancer (carcinogenic), may be responsible for birth defects (teratogenic) or, cause genetic alterations (mutagenic). The residues may affect soil fertility by eradicating the arthropod fauna and may affect neighbouring and successive crops. Koeman *et al*. (1972), in a survey of pollution of Lake Nakuru in Kenya, suggest that certain tropical regions where pesticides are extensively used they may be important sources for the observed atmospheric pollution.

Every year, FAO and WHO convene a joint meeting of experts to discuss the hazards to health arising from the contamination of foods with pesticide residues. The main findings and recommendations, including acceptable daily intakes, and tolerances for pesticide residues, are summarized in

a brief report. Often these include recommendations on controlled residues , i.e. toxicologically justified residues which are unavoidable by standards of good agricultural practice. They are quite distinct from 'poison in our food'. Summaries of the biological and chemical data on which recommendations are made are published in a brief report. For further information on the fate of insecticides in the environment, reference should be made to the volumes edited by Tahori (1971*a, b*). Some health aspects of pesticide residues are summarised by Bressau (1975).

10.8.4 Development of resistance

When an insecticide is used to control a pest, not all members of the population against which it is used are killed by the toxic material. Some, less susceptible to the particular insecticide at that dosage, survive. To control the surviving members, a higher concentration of insecticide would be required and a stage may be reached when the insecticide is totally ineffective against the pest: this is known as resistance. In fact, we are witnessing accelerated evolution as insect species adapt to new hazards, i.e. the insecticides, which have been introduced in their immediate environment. Resistant strains are derived from the initial population by the selective mortality of the more susceptible genotype following the application of insecticides (Sawicki, 1979). Resistance designates a genetic change, a response to pesticide selection.

According to Georghiou & Taylor (1977), the number of species of insects and acarines in which resistant strains have been reported has increased from 1 in 1908 to 364 in 1975 (Table 9). The impact of the development of resistance on modern pest control is very great indeed. Agriculturists are compelled to use higher dosages and more frequent applications to kill the same numbers of pests. This leads not only to greater disruption of the ecosystem than would otherwise happen when low dosages and less frequent applications were used but also to increased costs (in order to maintain present levels of agricultural productivity) and loss of investment in the development of insecticides. The number of available insecticides in the market is diminishing (Corbett, 1979). Some strains of insects and acarines have in turn developed resistance to arsenic, DDT and other chlorinated hydrocarbons, followed

Table 9 Number of species of insects and mites in which resistance to one or more chemicals has been documented (after Georghiou & Taylor, 1977).

Year	Species
1908	1
1928	5
1938	14
1948	15
1954	25
1957	76
1960	137
1963	157
1965	185
1967	224
1975	304 (+ 59 unconfirmed cases)

by organophosphates, carbamates and more recently by pyrethroids and all compounds available commercially for pest control. Sequential introduction of alternative pesticides, to deal with established resistance has led to wide-spread instances of cross-resistance (see later) and the withdrawal from use of alternative pesticides to control many notorious pests of agricultural and medical importance (Sawicki, 1975*a*). Recommended methods for the detection and measurement of resistance of agricultural pests to pesticides are detailed by Anonymous (1971*b*). Waterhouse (1977) provides discussion on standardized methods for the detection and measurement of resistance.

The development of resistance has been termed the most pressing problem of modern pest control. According to WHO (1976) 'resistance is probably the biggest single obstacle in the struggle against vector-borne diseases and is mainly responsible for preventing successful malaria eradication in many countries'. Resistance has been reported not only to the most recent insecticides but also to insect growth regulators, chemosterilants and even biological control agents (Sawicki, 1979).

Selection with insecticides Reports in the literature show that the development of effective resistance by an insect to an insecticide has not occurred until after the insect has been exposed to the chemical for several generations. Roussel & Glower (1957) and Graves *et al.* (1967) report that it took 25 generations of intensive selection for the boll weevil, *Anthonomus grandis*, to develop resistance to organochlorine compounds. DDT was used intensively for about 15 years before tobacco budworm, *Heliothis virescens*, developed resistance to this insecticide. According to Brown (1968), DDT resistance develops after an initial latent period of several generations before it increases steeply, whereas cyclodiene resistance develops without delay. Brown (1977), in discussing the over-all timescale of development of resistance (in terms of the numbers of species involved) to the main groups of insecticides states that the two types of organochlorine-resistance (DDT and cyclodiene) were developed usually in about 10 years, that for cyclodienes somewhat faster than the DDT group. He further states the resistance to organophosphorus compounds comes to fruition usually 10 years after the cyclodiene resistance. The resistance to carbamates takes the same time as the OP resistance but develops a little faster when it builds on a base of OP resistance. Other factors to be considered in selection with insecticides include the developmental stages of insects exposed, the type of insecticide application – whether residual or non-residual, etc.

Genetic basis The rate of development of resistance in an unexposed population is initially very low. But as the frequency of major genes for resistance in the surviving population is gradually increased, the insect becomes better organized genetically to exist in the contaminated environment. The more intense the selection pressure, the more rapid the development of resistance, provided that the number of survivors is large enough to maintain genetic variability (Georghiou, 1972). Further the seemingly limitless persistence of the resistance genes prevents the re-introduction of insecticides against populations which have apparently reverted to full susceptibility as a result of release of pressure from insecticides (Sawicki, 1979).

Ecological factors Several ecological factors may influence the development of resistance to insecticides in a population. This may depend on the relative isolation of populations from each other, by which the exchange of genetic material is influenced; variation of ecological time factors, such as season, may be important. Size, growth, generations per year and the reproductive potential of the populations may also influence the development of resistance.

Behavioural factors Several species of insects are known to have inherited ability to detect the presence of specific insecticides and escape before picking up lethal quantities. Examples are strains of mosquitoes more irritable to DDT deposits and strains of house-flies where avoidance of malathion increased with malathion resistance (Fay *et al.*, 1958).

10.8.5 Cross-resistance, multiple and multiplicate resistance

The term cross-resistance refers to resistance of a strain of insects to compounds other than the selecting agent, due to the same biochemical mechanism.

Multiple resistance occurs when two or more distinct mechanisms, each protecting against different poisons, are present together.

When two or more mechanisms co-exist in the same organism and protect it against the same poison, the animal is known to possess multiplicate resistance. Such a resistance usually results from the simultaneous or consecutive use of several insecticides under field conditions. In countries where many different insecticides have been used sequentially against house-flies, resistance may be both multiple and multiplicate (e.g. in Denmark). It is interesting to know that insects with a high level of resistance to chlorinated hydrocarbons may exhibit a high level of cross-resistance to carbamates (Graves *et al.*, 1967; Adkisson, 1968). Where resistance to an organophosphate is already present, exposure to an alternative organophosphate may result in more accelerated development of resistance to the new compound. This selection will also impart low levels of cross-resistance to various organophosphates; generally higher levels of resistance will occur only for the selecting compound and its close relatives. Sawicki (1979) reports that work at Rothamsted suggests that resistance to pyrethroids, particularly in certain peach potato aphid populations resistant to OPs and carbamates appears to result largely, *not from the prolonged or intensive use of pyrethroids*, but from cross-resistance conferred by resistance mechanisms selected through the use of other groups of insecticides. According to him 'pyrethroids are clearly victims of sequential resistance'.

10.8.6 Mechanism of resistance

How insect populations become resistant to a chemical is rarely known in detail. However, considerable information on the biochemistry, physiology and genetics of resistance has accumulated in the recent years, mostly on a few species of arthropods, and is summarized by Busvine (1971*b*), Oppenoorth (1975), Oppenoorth & Welling (1976), and Croft (1977). It should be emphasized that little or nothing is known about resistance mechanism in many agricultural pests. Sawicki (1979) classifies the identified resistance mechanisms into three broad categories: (*i*) delayed entry of the toxicant due to factors such as decreased permeability of the cuticle; (*ii*) decreased sensitivity of the site of action; and (*iii*) increased detoxification of the insecticide (Figure 16) or decreased activation by metabolic processes. The following points may be noted regarding the mechanism of resistance to some major insecticide groups:

Organochlorine compounds Whereas it was stated earlier that chlorinated insecticides are apparently toxic because of their effects on nervous system, there is evidence that they may also interfere with various metabolic processes in animals. One mechanism for DDT resistance was found early to be metabolic conversion to DDE (2, 2-bis-(p-chlorophenyl)-1, 1-dichloroethylene) by DDT dehydrochlorinase (Perry & Hoskins, 1950; Sternburg *et al.*, 1950). A gene responsible for this process, at least, in the house-fly, *Musca domestica*, is situated on chromosome II. Some resistant strains of insects, for example, pink bollworm, have increased detoxification of DDT while others exhibit decreased absorption of the compound.

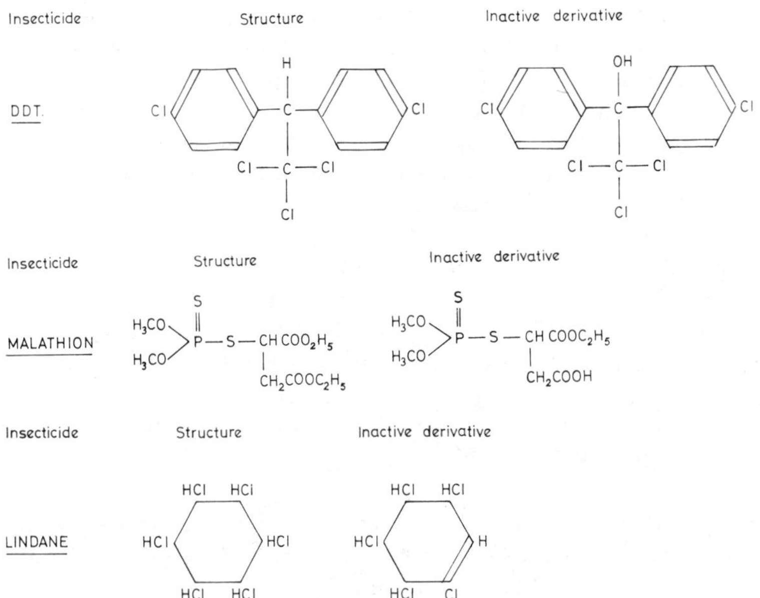

Figure 16 Examples of minor chemical modifications that render insecticide molecule less active.

Lindane is metabolized in insect tissues, but the mechanism is not clear. Dehydrochlorination is certainly one of the processes involved. In the house-fly a gene responsible for slower penetration of DDT is located on chromosome III which also carries the gene responsible for resistance to knockdown. The gene responsible for conversion of both DDT and DDE to the more easily excreted polar metabolities is situated on the 5th chromosome in house-flies. The genetics of house-fly resistance to DDT has been reviewed by Georghiou (1969), Sawicki (1975*b*) and Keiding (1977).

M. domestica also possesses other types of resistance against DDT but the bases of them are not yet known (see, for example, Busvine, 1971*b*). A knock-down resistance (kdr) in the house-fly is believed to be the result of decreased nerve sensitivity which causes resistance to DDT and the pyrethroids. Danish house-flies selected through wide-spread and prolonged use of DDT and possessing a kdr-like mechanism took only one season to develop high field resistance to pyrethroids (see

Sawicki (1975*a*) for further details). Plapp (1970), who has worked on the house-fly, has suggested that changes in ability to metabolize glucose may be a factor in some cases of insect resistance to chlorinated insecticides.

The mechanism of cyclodiene resistance is still not understood. Some workers agree that resistance to dieldrin in insects does not involve reduced penetration, increased metabolism or enhanced excretion of insecticides when compared with the same processes in dieldrin-susceptible insects. Schaefer & Sun (1967) believe that the dieldrin resistance in house-flies may be largely due to insensitivity at the neuromuscular receptor site. For an insight into the recent developments into mechanisms of resistance in insects to organochlorines, reference should be made to work of Oppenoorth & Welling (1976).

Organophosphorous compounds The biochemical mechanism for the toxicity of the OP compounds was first established with mammals. It is now known that metabolism of these compounds in mammals and insects is essentially the same (O'Brien, 1960), and Kilby (1965) concluded that the mammalian liver and the insect's fat body resemble one another biochemically in several respects. Although insects possess a variety of mechanisms resisting the toxic effects of insecticides, an important mechanism, depending on circumstances, is enhanced detoxification. OP poisoning in insects is also known to occur through the inhibition of acetylcholine esterase (O'Brien, 1967). However, in a few cases the property of resistance is due primarily to the possession of acetylcholinesterase which is insensitive to OP and carbamate insecticides. It was discovered by Apperson & Georghiou (1975) that slower penetration and increased metabolism of the chemical was responsible for resistance of OP compounds to larvae of *Culex tarsalis*, and not the insensitivity of acetylcholinesterase to organophosphates. Sawicki (1979) believes that at least three factors control or modify resistance to OPs: decreased penetration through the cuticle, detoxification mechanisms and a decrease in the sensitivity of the cholinesterase. Of these, increased detoxification is now believed to be the principal cause of resistance, in insects resistant to OPs.

Genetic studies show that resistance to OP compounds is highly complicated though it may, sometimes, be caused by a single gene in any given strain. In *Drosophila melanogaster*, the main OP resistance gene, called 'a' because of its association with reduced aliesterase activity is located on chromosome IV (Brown, 1968). Three different detoxifying mechanisms in the parathion-resistant house-flies are believed to be controlled by different genes on chromosome II (see Sawicki, 1975*b* for references). Delayed penetration in many resistant strains of house-flies is believed to be controlled by gene Pen on Chromosome III though the nature of Pen itself is unknown. A detailed recent article on the genetics of resistance of house-flies to insecticides is by Sawicki (1975*b*).

Carbamates Metcalf *et al*. (1968) considered the insecticidal carbamates to be active anticholine esterases which, as stated earlier, cause characteristic symptoms of cholinergic stimulation similar to those caused by the OP insecticides. Recently, evidence has been accumulating that in house-flies, oxidative mechanisms may be responsible for resistance. The occurrence of high levels of oxidases in the microsomes of carbamate resistant house-flies (Tsukamoto & Casida, 1967), as noted by Plapp (1970), supports this theory. Resistance to carbamate insecticides in the house-fly, *M. domestica*, has sometimes been attributed to a single gene and sometimes to polygenic systems. Plapp (1970) has concluded that in the house-fly, resistance to carbamate insecticides is likely to be due to the interaction of genes located on chromosomes II, III, and V. According to Brown (1968), in whichever insect investigated the OP resistance gene has always proved to be dominant, as has that for carbamate-resistance. In some cases DDT has been found to be recessive, but not in the house-fly; dieldrin-resistance is always semi-dominant.

10.9 Future use of insecticides

Clearly one approach to the control of pests would be to develop chemicals which are highly selective and safe (see Gasser, 1966 for discussion). The articles by Elliott (1977, 1979) and Elliott and Janes (1979) suggest the possibility of synthesizing a wide range of compounds with a combination of useful characters such as high insecticidal activity, low mammalian toxicity, and controlled environmental stability. Unfortunately, as discussed earlier, the development of insecticides is a costly operation. Corbett (1979) states that there is evidence to support the view that at the moment the pesticide industry is no longer in a rapid growth phase of technical innovation. According to him the current frequency of invention is 16 products per annum, while the number for the late 1960s was 25 per annum. Earlier, Lewis (1977) while examining the pattern of introduction of new insecticides concluded that post-1970 has witnessed a precipitous fall in the annual introduction of pesticides. If this trend continues very soon few new chemicals will be available for pest control. The second approach is the controlled use of existing pesticides, as detailed below. In this way, pesticides can help to maintain the ecological equilibrium in our environment while providing optimum benefits. It must be recognized that to use pesticides is as much a calculated risk as the decision *not* to use them when required, and adherence to the following rules may assist in achieving the goals sought by entomologists (pesticide usage in Nigeria has been discussed in the symposium edited by Youdeowei & Fadare, 1975).

1) Use pesticides only when the need is clearly established. For this purpose determine accurately economic threshold infestation levels for each pest.
2) Reduce the amount of pesticides used in the field by the development of reliable pest prediction methods. In this way pesticides can be used when most effective.
3) Determine when biological control agents of the pests are most active in the field and avoid chemical treatments during this period.
4) Where possible combine insecticides with other non-chemical methods in integrated pest management systems. For example, very promising results have been obtained when chemicals have been used in conjunction with cultural practices. Attiah (1977) reports from Egypt that by planting corn not later than mid-June, the areas to be sprayed with insecticide decreased to an average of of about 9000 ha during the seasons 1970–75 whereas it had reached 280 000 ha during the 1964 season before

this procedure was implemented. Similarly, even partially resistant varieties of crops may drastically curtail the need of insecticides.

5) The problem of development of pest resistance must be handled very carefully and the following points may be noted in connection with measures for delaying, reducing or avoiding development of resistance:

(*i*) It is easier to stop an emerging resistance to pesticides than to reverse when fully developed. The nature of the resistance, whether it is behavioural or has a metabolic basis must be determined and appropriate steps taken to deal with the situation.

(*ii*) Where feasible, use specific pesticides for they are not easily bypassed by metabolic adjustment.

(*iii*) Use only the dosage sufficient to achieve control of economic losses. Overdose does not bring eradication but it renders selection unnecessarily effective. Increased dosages can be used later to overcome slight resistance. Never use insecticide treatments as an insurance policy, i.e. on a prophylactic basis.

(*iv*) Combine pesticides with baits, strips etc. to achieve high concentration on the targets with low contamination of the environment.

(*v*) Spare reservoirs of low pest density for immigration of non-resistant immigrants into a treated area. Reversion of resistance depends on existing non-resistant alleles! Combine insecticides with non-chemical methods in an integrated pest management scheme.

(*vi*) Combine insecticides with synergists, i.e. inhibitors of enzymes which degrade pesticides in the insect body, e.g. piperonyl butoxide + pyrethrum. Evidence is emerging that alternative chemicals, including juvenile hormones, chemosterilants and synergists do not avoid the danger of resistance. In any case, it is clear that present day insecticides will be an essential element in most programmes of integrated pest management. Thus the question of resistance should no doubt continue to remain central in pest control strategies. However, much remains to be learned regarding the genetics and biochemistry of resistance of insects to insecticides. Apart from large gaps in our knowledge of these fields, a major unknown is the rate at which susceptibility to a class of insecticide reappears when a specific insecticide pressure is totally removed and when other classes of insecticides are used. Nor is it usually known how quickly resistance develops when using sequences of different insecticides. With this knowledge it should be possible to develop a spraying regime which greatly delayed, or perhaps completely avoided the development of resistance to insecticides.

10.10 Literature cited

Adeyemi, S. A. O. (1970). Handbook of agricultural insecticides available in Nigeria. *The Ministry of Agriculture and Natural Resources, Western State of Nigeria and the Entomological Society of Nigeria*. 56 pp.

Adkisson, P. L. (1968). Development of resistance, by the tobacco budworm to endrin and carbaryl. *J. Econ. Ent.*, **61**, 37–40.

Apperson, C. S. & Georghiou, G. P. (1975). Mechanisms of resistance to organophosphorous insecticides in *Culex tarsalis, J. Econ. Ent.*, **68(2)**, 153–7.

Anonymous (1971*a*). The story of modern agrochemicals. *PANS*, **17(3)**, 304–7.

Anonymous (1971*b*). Recommended methods for the detection and measurement of resistance of agricultural pests to pesticides. FAO method no. 7. *FAO Plant Protection Bulletin*, **19**, 15–18.

Anonymous (1972). *Pesticides in the Modern World*. A symposium prepared by members of the Co-operative Programme of Agro-Allied Industries with FAO and other United Nations Organisations, 1972. Printed by Newgate Press Ltd., London. 59 pp.

Anonymous (1977). Pyrethroids. In *Notes and News. PANS*, **23(3)**, 335.

Attiah, H. H. (1977). Ecological assessment of pesticide management on terrestrial ecosystem in Egypt. *Proc. UC/AID-UNIV. Alexandria, A.R.E. Seminar/Workshop in pesticide management*. 61–74.

Braunholtz, J. T. (1968). Organo-phosphorous insecticides. *PANS*, **14(4)**, 467–84.

Braunholtz, J. T. (1977). Pesticide development and the chemical manufacturer. *Proc. XV. International Congr. Ent. Washington*, 747–55.

Braunholtz, J. T. (1979). Techno-economic considerations. *Chemistry and Industry*, 17 November issue, 789–91.

Breese, M. H. & Searle, R. J. C. (1977). Why the newer synthetic pyrethroids show promise. *SPAN*, **20**, 18–20.

Bressau, G. (1975). Health aspects of pesticides and their residues. *EPPO Bull.*, **5(2)**, 73–8.

Brooks, G. T. (1974). *Chlorinated Insecticides.* Cleveland, Ohio. CRC Press (Blackwells, London). Vol. I (249 pp.), Vol. II (197 pp.).

Brown, A. W. A. (1951). *Insect Control by Chemicals*. John Wiley, New York. 817 pp.

Brown, A. W. A. (1960). Mechanisms of resistance against insecticides. *Ann. Rev. Ent.*, **5**, 301–26.

Brown, A. W. A. (1968). Insecticide resistance comes of age. *Bull. Ent. Soc. Amer.*, **14(1)**, 3–9.

Brown, A. W. A. (1977). Epilogue: Resistance as a factor in pesticide management. *Proc. XV. International Congr. Ent. Washington,* 816–24.

Busvine, J. R. (1971*a*). A critical review of the techniques for testing insecticides. *Commonwealth Agricultural Bureaux, England*. 395 pp.

Busvine, J. R. (1971*b*). The biochemical and genetic bases of insecticide resistance. *PANS*, **17(2)**, 135–46.

Carson, R. (1962). *Silent Spring*. Houghton Mifflin Co., Boston. 368 pp.

Casida, J. E. (Ed.) (1973). *Pyrethrum: the natural insecticide*. Academic Press Inc., London and New York. 329 pp.

Coaker, T. H. (1977). Crop pest problems resulting from chemical control, p. 313–28. In *Origins of Pest, Parasite, Disease and Weed Problems*. 18th symposium of the British Ecological Society, Bangor, 12–14 April, 1976. Eds J. M. Cherret and G. R. Sagar. Blackwell Scientific Publications, Oxford, 413 pp.

Corbett, J. R. (1979). Technical considerations affecting the discovery of new insecticides. *Chemistry and Industry*, 17 November issue, 772–82.

Croft, B. A. (1977). Resistance in arthropod predators and parasites. pp. 377–93. In *Pesticide Management and Insecticide Resistance.* Eds D. L. Watson and A. W. A. Brown. Academic Press, New York. 638 pp.

Cunningham, J. H. (1973). Pesticides and the economy of the developing world. *SPAN*, **16(2)**, 54–5.

Davies, J. H. (1968). Developments in non-organophosphorous insecticides. *PANS*, **14(4)**, 485–504.

DeBach, P. & Bartlett, B. (1951). Effects of insecticides on biological control of pests of citrus. *J. Econ. Ent.*, **44**, 372–83.

DuBois, K. P. (1963). Toxicological evaluation of the anticholinesterase agents, p. 833–59. In *Cholinesterase and Anticholinesterase Agents*. Ed. G. B. Loelle., Springer Verlag, Berlin. 1220 pp.

Elliott, M. (1977*a*). Synthetic insecticides designed from the natural pyrethrins. *Pontificiae Academiae Scientiarvm Scripta Varia*, **41**, 157–76. (Published as: Marini-Bettólo, G. B. (Ed.) (1977). *Natural Products and the Protection of Plants*. Proceedings of a study week at the Pontifical Academy of Sciences. October 18–23. Elsevier Scientific Publishing Company, Amsterdam. 846 pp.)

Elliott, M. (1977*b*). *Synthetic Pyrethroids.* ACS symposium series No. 42, American Chemical Society, Washington, D. C. 229 pp.

Elliott, M. (1979). Progress in the design of insecticides. Holroyd Memorial Lecture. *Chemistry and Industry*, 17 November issue, 757–68.

Elliott, M., Farnham, A. W., Janes, N. F., Needham, P. H. & Pearson, B. C. (1967). 5-Benzyl-3 furylmethyl chrysanthemate: a new potent insecticide. *Nature,* **213**, 493–4.

Elliott, M., Farnham, A. W., Janes, N. F., Needham, P. H., Pulman, D. A. & Stevenson, J. H. (1973), NRDC 143, a more stable pyrethroid. *Proc. 7th Brit. Insect. & Fung. Conf. Brighton*, 721–8.

Elliott, M., Farnham, A. W., Janes, N. F., Needham, P. H. & Pearson, B. C. (1974). Synthetic insecticides with a new order of activity. *Nature,* **248**, 710-11.

Elliott, M. & Janes, N. F. (1979). Synthetic pyrethroids – a new class of insecticides. *Chem. Soc. Rev*., **7**, 473–505.

Eto, M. (1974). *Organophosphorous Pesticides: organic and biological chemistry*. CRCC Press, UniScience Series. Cleveland, Ohio. 375 pp.

Fletcher, W. (1974). *The Pest War.* Basil Blackwell, Oxford. 176 pp.

Fay, R. W., Kilpatrick, J. W. & Morris, G. C. (1958). Malathion resistance studies on the housefly. *J. Econ. Ent*., **51**, 452–3.

Gasser, R. (1966). Use of pesticides in selective manners. *Proceedings of the FAO symposium on integrated pest control*, 11–15 October, 1965. Rome 109–13.

Georghiou, G. P. (1969). Genetics of resistance to insecticides in houseflies and mosquitoes. *Exp. Parasitol*., **26**, 225–55.

Georghiou, G. P. (1972). The evolution of resistance to pesticides. *Ann. Rev. Ecology and Systematics*, **3**, 133–68.

Georghiou, G. P. & Taylor, C. E. (1977). Pesticides resistance as an evolutionary phenomenon. *Proc. XV International Congr. Ent. Washington*, 759–85.

Gilbert, C. H. (1978). The increasing riskiness of the pesticide business. *Farm Chemicals.* **141** (April), 20, 22, 24, 26–7.

Graves, J. B., Roussel, J. S., Gibbens, J. & Patton, D. (1967). Laboratory studies on the development of resistance and cross-resistance in the boll weevil. *J. Econ. Ent*., **60**, 47–50.

Gunther, F. A. & Jeppson, L. R. (1960). *Modern Insecticides and World Food Production*. Chapman and Hall, London. 284 pp.

Hickey, J. J., Keith, J. A. & Coon, F. B. (1966). An exploration of pesticides in a Lake Michigan Estuary. *J. appl. Ecol*., **3**, (suppl.), 141–54.

Highwood, D. (1980). Unexpected bonuses from the pyrethroids. *Shell in Agriculture,* February issue, 5–6.

Keiding, J. (1977). Resistance in the housefly in Denmark and elsewhere, pp. 261–302. In *Pesticide Management and Insecticide Resistance.* Eds D. L. Watson and A. W. A. Brown. Academic Press, New YOrk. 638 pp.

Kilby, B. A. (1965). Intermediary metabolism and the insect fat body, pp. 39–48. In *Aspects of Insect Biochemistry*. Ed. T. W. Godwin. *Biochem. Soc. Symp.,* **25**, 1–107.

Koeman, J. H., Pennings, J. H., DeGoeij, J. J. M., Tjioe, P. S., Olindo, P. M. & Hopcrafts, J. (1972). A preliminary survey of the possible contamination of Lake Nakuru in Kenya with some metals and chlorinated hydrocarbon pesticides. *J. appl. Ecol.*, **9(2)**, 411–16.

Kuhr, R. J.& Dorough, H. W. (1976). *Carbamate Insecticides: chemistry, biochemistry and toxicology*. CRC Press, Uniscience Series. Cleveland, Ohio. 300 pp.

Kumar, R. (1979). Biological control of cocoa pests in West Africa. *Proc. 5th Conf. West African Cocoa Entomologists. Yaounde*, 1976, 109–16.

Leatherdale, D. (1966). The use of systemic insecticides in warm climates. *PANS*, **12(4)**, 288–308.

Lewis, C. J. (1977). The economics of pesticide research, pp. 237–45. In *Origins of Pest, Parasite, Disease and Weed Problems*. 18th symposium of the British Ecological Society, Bangor, 12–14 April, 1976. Eds J. M. Cherret and G. R. Sagar. Blackwell Scientific Publications. Oxford. 413 pp.

Lincer, J. L., Zalkind, D., Brown, L. H., Hopcraft, J. (1981). Organochlorine residues in Kenya's rift valley lakes. *J. appl. Ecol.*, **18(1)**, 157–71.

Matsumura, F. (1975). *Toxicology of Insecticides*. Plenum Publishing Corporation, New York. 504 pp.

Mercer, S. L. (1981). *An Index of Chemical, Common and Trade Names of Pesticides*. Tropical Pest Management Pesticide Index – 1981. 61 pp.

Metcalf, R. L. (1955). *Organic Insecticides*. Interscience, New York. 392 pp.

Metcalf, R. L. (1971). Structure-activity, relationships for insecticidal carbamates. *Bull. Wld. Hlth. Org.*, **44**, 43–78.

Metcalf, R. L. (1980). Changing role of insecticides in crop protection. *Ann. Rev. Ent.*, **25**, 219–56.

Metcalf, R. L., Gruhn, W. G. & Futuko, T. R. (1968). Electrophysiological action of carbamate insecticides in the central nervous system of the American cockroach. *Ann. Ent. Soc. Amer.*, **61**, 618–24.

Metcalf, R. L. & McKelvey, J. J. (Eds). (1976). The future for insecticides – needs and prospects. *Adv. Env. Sci. & Tech. Ser.*, Vol. **6**. John Wiley, New York. 432 pp.

Moriarty, F. (Ed.) (1975). *Organochlorine Insecticides: persistent organic pollutants*. Academic Press, London. 302 pp.

NAS (National Academy of Sciences) (1975). *Pest Control: an assessment of present and alternative technologies*. Vol. 1. Contemporary pest control practices and prospects: the report of the executive committee, Washington, D.C. 506 pp.

Newsom, L. D. (1967). Consequences of insecticide use on non-target organisms. *Ann. Rev. Ent.*, **12**, 257–86.

O'Brien, R. D. (1960). *Toxic Phosphorus Esters – chemistry, metabolism and biological effects*. Academic Press, New York. 434 pp.

O'Brien, R. D. (1967). *Insecticide Action and Metabolism*. Academic Press, New York. 332 pp.

O'Brien, R. D. (1978). The biochemistry of toxic action of insecticides, pp. 515–39. In *Biochemistry of Insects.* Ed. M. Rockstein. Academic Press, New York. 649 pp.

Oppenoorth, F. J. (1975). Biochemistry and physiology of resistance, WHO. VBC/75, **5**, 1–40.

Oppenoorth, F. J. & Welling, W. (1976). Biochemistry and physiology of resistance, pp. 507–51. In *Insecticide Biochemistry and Physiology.* Ed. C. F. Wilkinson. Plenum, New York. 768 pp.

Owusu-Manu, E. (1976). Natural enemies of *Bathycoelia thalassina* (Herrich-Schaeffer), a pest of cocoa in Ghana. *Biol. J. Linn. Soc.*, **8(3)**, 217–44.

Perring, P. H. & Mellanby, K. (1978). *Ecological Effects of Pesticides.* Academic Press, London. 194 pp.

Perry, A. S. & Hoskins, W. M. (1950). The detoxification of DDT by resistant houseflies and inhibition of this process by piperonyl cyclonene. *Science*, **111**, 600–1.

Pimentel, D., Krummel, J., Gallahan, D., Hough, J., Merrill, A., Schreiner, I., Vittum, P., Koziol, F., Back, E., Yen, D. & Fiance, S. (1978). Benefits and costs of pesticide use in U.S. food production. *BioScience,* **28(12)**, 772–84.

Plapp, F. W. Jr. (1970). Inheritance of dominant factors for resistance to carbamate insecticides in the housefly. *J. Econ. Ent.*, **63(1)**, 138–41.

Ripper, W. E. (1956). Effect of pesticides on balance of arthropod populations. *Ann. Rev. Ent.*, **1**, 403–36.

Roussel, J. S. & Glower, D. F. (1957). Resistance to the chlorinated hydrocarbon insecticides in the boll weevil. *J. Econ. Ent.*, **50**, 463–8.

Sawicki, R. M. (1975*a*). Effects of sequential resistance on pesticide management. *Proc. 8th Brit. Insect. & Fung. Conf.*, 799–811.

Sawicki, R. M. (1975*b*). Some aspects of the genetics and biochemistry of resistance of house-flies to insecticides. *WHO: Expert committee on resistance of vectors and reservoirs to pesticides. VBC/EC/75,* **10**, 1–13.

Sawicki, R. M. (1979). Resistance to pesticides 1. Resistance of insects to insecticides. *SPAN*, **22(2)**, 50–2, 87.

Schaeffer, C. H. & Sun, Y. P. (1967). A study of dieldrin in the housefly central nervous system in relation to dieldrin resistance. *J. Econ. Ent.,* **60(6)**, 1580–3.

Sternburg, J. C., Kearns, W. & Bruce, W. N. (1950). Absorption and metabolism of DDT by resistant and susceptible houseflies. *J. Econ. Ent.,* **43**, 214–19.

Tahori, A. S. (Ed.) (1971*a*). Pesticide chemistry. Volume VI. Fate of pesticides in environment. *Proc. 2nd International Congr. Pesticide Chemistry.* Gordon & Breach Scientific Publishers Inc., London. 571 pp.

Tahori, A. S. (Ed.) (1971*b*). *Pesticide Terminal Residues*. Invited papers from the international symposium on pesticide terminal residues, Tel-Aviv, Israel. Butterworths Scientific Publications, London. 365 pp.

Tahori, A. S. & Galun, R. (1976). The rise and fall of DDT. *Israel Journal of Entomology*, **11**, 33–51.

Tsukamoto, M. & Casida, J. E. (1967). Metabolism of methylcarbamate insecticides by the $NADPH_2$-requiring enzyme system from houseflies. *Nature*, **213**, 49–51.

Waterhouse, D. F. (1977). FAO activities in the field of pesticide resistance. *Proc. XV. International Congr. Ent. Washington,* 786–93.

Weiden, M. H. J. & Moorefield, H. H. (1965). Synergism and species specificity of carbamate insecticides. *J. Agric. Food. Chem.,* **13**, 200–4.

Whitney, W. K. (1969). The discovery and development of new pesticides. *Bull. ent. Soc. Nigeria*, **2(1)**, 82–4.

WHO (World Health Organization) (1976). Resistance of vectors and reservoirs of disease to pesticides. *WHO, Geneva. 22nd Rpt. of Ser.,* **585**, 88 pp.

Wilkinson, C. F. (Ed.) (1976). *Insecticide Biochemistry and Physiology.* Plenum Press, New York. 768 pp.

Worthing, C. R. (1979). *The Pesticide Manual*. British Crop Protection Council, London. 6th Edition. 655 pp.

Youdeowei, A. & Fadare, T. A. (1975). Pesticide usage in Nigeria. Proceedings of a symposium held at Ahmadu Bello University, Zaria. *Ent. Soc. Nig. Occ. Publications,* **17**, 9–99.

11 Formulation and Application of Insecticides

11.1 Formulation

Insecticides are, in general, used in extremely small dosages. The active ingredient(s) of a pesticide formula must be distributed over a wide area as evenly as possible. So these chemicals must be formulated. Formulation is the science or art of diluting the insecticide evenly in a suitable manner for application through appropriate machinery, conveniently, efficiently, safely and at the same time in forms which are toxic against the pests involved. The type of formulation to be used will be governed by a number of parameters including the nature of the insecticide itself, its mode of action, the ecology of the pest to be controlled and the available method of application. The last will be governed by the economics of the agricultural system where the formulation is to be used. The quality and sophistication of the formulation used can have a marked effect on the performance of the pesticide under field conditions. Technical assistance and consultations from international organizations such as UNIDO and FAO are now available to developing countries for the establishment of local pesticide production and formulation industries. A variety of formulations especially of liquid formulations are available. The common types now in use are discussed below. For further details of various aspects of formulations, see Bals (1973), Furmidge & Shenton (1973), Marrs & Middleton (1973), Van Valkenburg (1973) and Matthews (1979). The dynamics of applied pesticide in the local environment in relation to biological response is discussed by Hartley & Graham-Bryce (1980).

11.1.1 Dusts

Pesticides were first applied as dusts. The active ingredient is mixed with an inert substance such as talcum powder, or with some other finely-divided material. The mixture must flow easily without forming lumps and should be compatible with the insecticide. Dusts are best applied either when the crop is wet, or when dew is on the plants so that the dust will easily stick to the plant. Dusts are used to protect stored produce and provide seed treatment. Compared with other formulations the advantages of dusts are as follows:

(1) Dusts may be bought ready to use so no further dilution is required and unlike wettable powders or emulsions, no water is needed.

(2) They are comparatively less phytotoxic than some other formulations. Although less likely to be absorbed by the user through the skin, there is a risk that small particles will be inhaled. The main disadvantages of dusts are the following:

  (*i*) They are highly susceptible to drift during applications and are readily removed from plant surfaces by wind and rain. Thus the residual life of dusts is usually shorter than other formulations.

  (*ii*) They are bulky to transport because much inert material must be conveyed from the manufacturer to the user. They are messy to handle and are not always effective in action, especially the stability of low concentration dusts is often poor.

11.1.2 Granules

Dusts have now been largely replaced by formulations containing larger particles of inert material. Granular formulations are termed as *preformed* if they are produced by absorbing a chemical onto a granular carrier or by coating an inert core (Marrs & Middleton, 1973). The formulation is called agglomerated if prepared by agglomeration of a powdered mixture of inert filler and insecticide. In *extruded* formulations a paste of toxicant, filler and binder is forced through small orifices. The size of granules varies from 0.01–3 mm in diameter, the usual range being between 0.3–0.7 mm. Microgranules of about 0.01–0.02 mm diameter are designed to adhere to foliage. Actually the granule size affects placement, distribution and release rate of the insecticide. Furmidge *et al.*

(1966) and Furmidge (1972) identify formulation type, application method and environmental condition as the major factors affecting the performance of granules. Granules are highly suitable against soil-inhabiting pests and for distribution of systemic chemicals which translocate through the roots. They have been used with success in flooded rice fields against stem-borers and in public health for controlling mosquito larvae, however, they need more care when applied by hand. In Africa granular formulations have been used for control of maize stemborers. Substantial improvement in granular formulations may be possible to achieve with the use of systemic insecticides by using controlled release formulations. The main benefits of granular formulations are as follows:

(1) Granules are easy to apply by hand-scattering (preferably using a gloved hand), thus obviating the need of expensive machinery.
(2) Because of their size, granular formulations do not drift and can be kept on site.
(3) Use of controlled-release granules reduces the rate of breakdown of an insecticide by factors such as hydrolysis or biodegradation and hence prolongs exposure. Losses due to leaching and volatilization are also minimized.
(4) Granular insecticides are generally known to be less harmful to natural enemies of pests and bees (Free *et al.*, 1967).

11.1.3 Wettable powders

Wettable (or dispersible) powders usually consist of a mixture of water-insoluble active ingredient insecticide, a dispersing agent, a wetting agent and a filler such as china clay. As noted by Marrs & Middleton (1973) when the product is added to water, the wetting agent enables the particle to be 'wetted' and the dispersing agent prevents the particles from aggregating together. Wettable powders contain a high proportion of active ingredient. SEVIN 85 WP, for example, contains 85% of the insecticide carbaryl. Wettable powders are easily manufactured and being relatively cheaper than emulsifiable concentrates, constitute perhaps one of the most widely-used spray formulations. They have the disadvantage of having to measure out a powder which can puff up into operator's face. Further, they blow about in wind and contaminate the environment.

Wettable powders are usually as effective if not more than emulsifiable concentrates when crawling insects are the target due to surface particulate deposit. As with dusts, their main advantage over liquid formulation is their cheapness. In Africa, some wettable powders have been prepackaged in sachets ready for individual knapsack loads.

11.1.4 Emulsifiable concentrates

An emulsifiable concentrate is usually a solution of the insecticide in a solvent such as xylene mixed with an emulsifier so that when mixed with water, an emulsion is formed. This is a dispersion of globules of the insecticide in the continuous medium of water. An emulsifiable concentrate is more expensive due to the solvents needed. They are inflammable and may deteriorate during storage at extreme temperatures so should be kept in a shaded shed. The proportion of active ingredient in an emulsion is higher than that found in dusts, but lower than that of wettable powders.

11.1.5 Aerosols, fogs, fumigants and smokes

An *aerosol* is a spray with droplets smaller than 50 μm which remain airborne, so best results are obtained when used in enclosed spaces. An aerosol can be obtained from a pressure pack by forcing an insecticide solution through a jet with a propellent gas. *Fogs* (droplets usually less than 15 μm) are produced by condensing a vapour produced when an insecticide in an oil solution is fed into a hot gas. A *smoke* may be obtained by burning an insecticide mixed with a suitable combustible material. *Fumigants* are liquids with comparatively high vapour pressure and may either be used in soil or in enclosed spaces. They are often the only effective means of controlling pests of stored products. In Africa cocoa beans are compulsorily fumigated at the harbours before shipment for exports. They (fumigants) act by penetrating the insect's respiratory system but may also be absorbed through the cuticle. Fumigants have the advantage of penetration, accurate dosage, short application time, 100% toxic action and rapid dispersal. They however, often require costly equipment and skilled supervision and must have no residual action. The use of fumigants is discussed by Page & Lubatti (1963).

11.1.6 Seed treatment

Proper seed treatment with insecticides is important in some cases to ensure a satisfactory plant stand and protect the young seedlings. Seed treatment is often considered an economic insurance policy and is a comparatively less expensive method of insecticide application. Seeds treated with persistent organochlorines are however, known to cause mortality among seed-feeding birds. Seed dressings may be liquids absorbed by the seed coats or a mixture of an insecticide with a 'sticker' enabling the active ingredient to stick directly to the seed coat. Pelleted seeds incorporating the insecticide and an inert substance such as clay have also been marketed but high costs have limited their use.

11.1.7 Encapsulation

Surrounding the pesticide with a protective shell or coating, for controlled release later is termed encapsulation (Marrs & Middleton, 1973). The protective shell is generally an envelope of material non-toxic by contact but toxic to insects ingesting it. It is effective as a stomach poison only and thus specific against insects with biting mouthparts (Phillips, 1968). As noted by Marrs & Middleton (1973), capsules can now be produced with sizes ranging from 1–2 mm (granular size) down to a few microns (microcapsules) and be formulated as powders, granules or suspensions. A major advantage with the encapsulation method is that because of protection of the encapsulated material from degradation by the environment, the rate of release of the chemical, and hence its persistence, in the field can be controlled. Handling of the toxic active ingredient is reduced provided capsules do not break. Spray teams can use conventional spray equipment to apply encapsulated formulations. The repellent effect of odour or flavours of an insecticide formulation is masked, natural enemies are usually less affected (except bees which might collect the pollen-sized capsules) and formulations can be 'tailor-made' to suit a particular situation and to give the formulation the required specificity or persistence (Phillips, 1968). Encapsulated formulations are relatively expensive and hence have not been widely used so far. For much useful information on encapsulation, reference should be made to the papers by Phillips (1968) and Marrs & Middleton (1973).

11.1.8 Baits

This involves the use of suitable attractants to lure the pest to the insecticide. For example, in Africa, bran or any substance that locusts will eat and is cheaply available locally is mixed with an insecticide such as Gamma-BHC and is distributed thinly over the locusts, though this method is now less used due to cost of transporting tonnes of bait. Use of ULV sprays on vegetation is preferred. A variety of materials that can be used in this way are now available. They seem specially suitable for use by small scale farmers. 'Granules' of citrus pulp etc. have been used as bait for leaf-cutting ants etc., baits of protein hydrolysate for attracting fruit-flies and addition of molassess to attract moths. The use of pesticides with baits, strips etc. to achieve high concentration on the targets with low contamination of the environment is currently gaining importance.

11.2 Application of insecticides

The choice of equipment for the application of insecticides must be consistent with the recommended method of control and will depend on the physical nature of the formulation of the pesticide to be used. For optimal chemical effect, it is essential that the pesticide is evenly and continuously dispersed over the target area during the operation. A wide range of spraying equipment is available on the market. Deutsch (1976) lists sprayers, dusters and other application equipment manufactured by 400 firms in 26 countries. Anonymous (1975) provides useful information on a variety of pesticide application equipment. An insight into the spraying systems for the 1980s is provided in the book edited by Walker (1980). The small scale African farmer needs relatively simple equipment suitable for the range of crops in his farming system. This current trend towards decreasing the total volume of application, and thus reduce the cost and effort of carting diluent (usually water) and time required for application is especially suitable for African farmers. On small farms, under 5 ha, hand-operated machinery is usually adequate but for larger acreages, power operated sprayers may be needed. In Uganda hand operated knapsack sprayers have been used on cotton grown areas up to

50 ha. As noted by Tunstall & Matthews (1972), an animal drawn sprayer is technically intermediate between the hand-operated knapsack sprayer and completely mechanized spraying and is suited to a wide range of conditions. It is practicable where draught animals are in use for other farming operations.

In the African countries, where labour is readily available, the manoeuvrability and low initial costs of small equipment is an advantage over large tractor or animal-drawn equipment. Indeed, under difficult African conditions, where road access to farms rarely exists and tractors are beyond the reach of the farmer, a hand operated knapsack sprayer that can be carried on a man's back has been the most widely-used means of applying insecticide. Further, under conditions such as those in the cocoa farm, it would be impossible to use tractor drawn equipment. Tractors are not suitable for much of Africa due to problems of cost, maintenance and soil conditions.

11.2.1 Sprayers

Most sprayers, while differing in very many ways have the following essential parts (Fig 17).

(1) Container for the insecticide to be sprayed.
(2) A gauze filter to keep dirt out of the pump.
(3) A pump for creating pressure inside the container to force the chemical spray out.
(4) A flexible pipe or hose to receive the chemical forced by the pump.
(5) An extension pipe sometimes called the lance or boom to carry single or multiple nozzles.
(6) One or several nozzles which create the spray.

Nozzles Different types of nozzles are used to create different spray patterns. For example, for spraying the ground or wall surfaces a 'fan' nozzle is best as it produces a flat fan of spray (Fig. 18**a**). For general spraying of foliage the cone or swirl nozzle (Fig. 18**b**), which produces a cone of spray, is used. A small hole with high pressure gives a fine mist-like spray. The cone or swirl nozzle has two discs separated by a washer to form a swirl chamber. The inner disc has slanting holes so that the liquid forced through them is made to swirl. The outer disc has a central hole; discs with different size holes are available. Droplet size depends on the design of the nozzle and pressure.

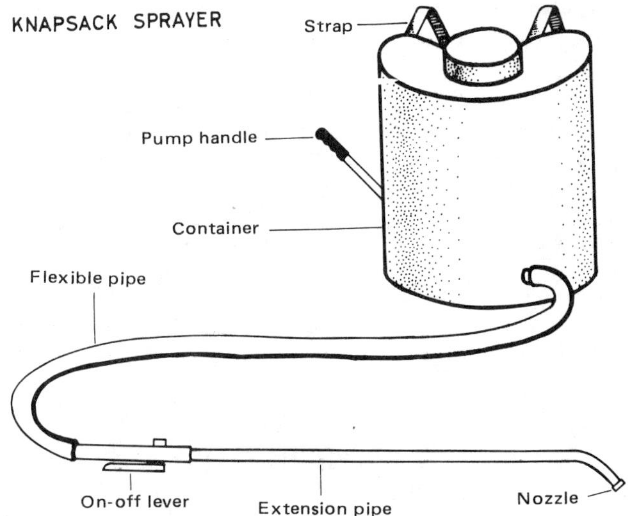

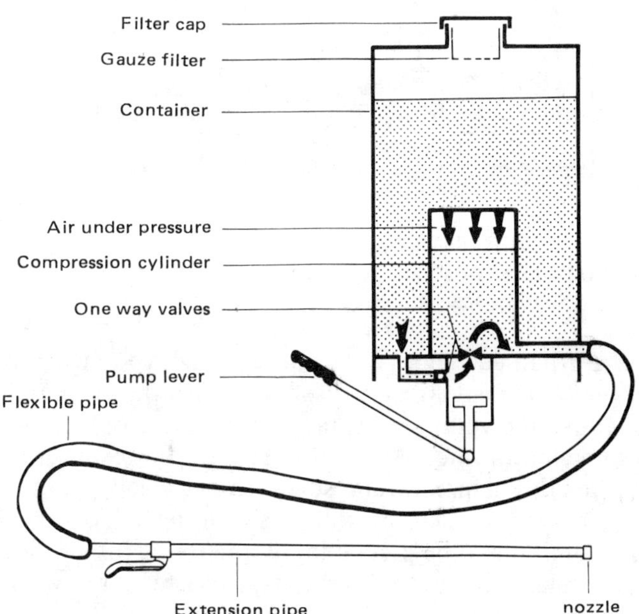

Figure 17 (**a**) Knapsack sprayer. (**b**) Parts of a sprayer.

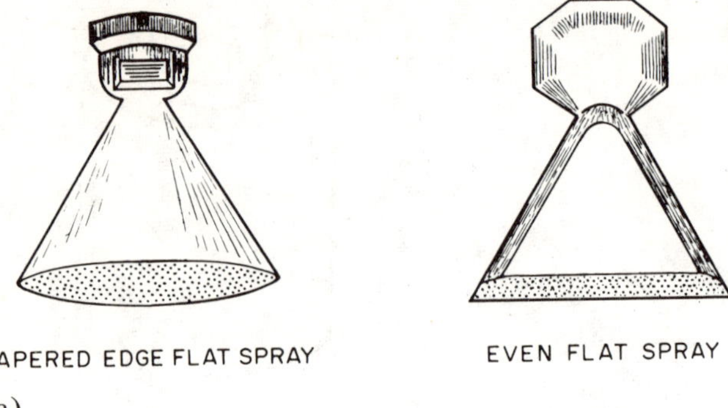

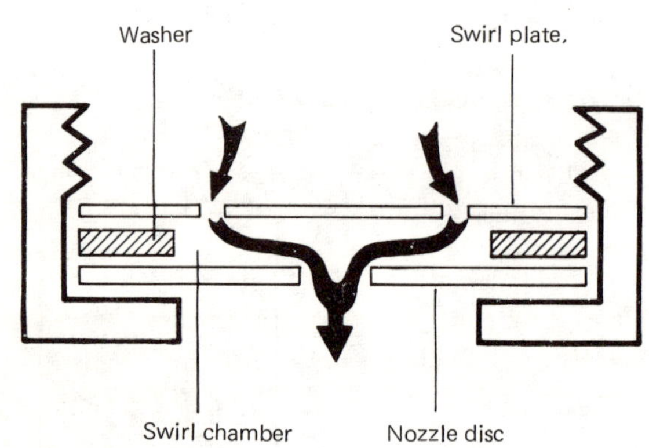

Figure 18 (**a**) Patterns of a 'fan' nozzle spray.
(**b**) Swirl nozzle.

Droplet size When pesticides are applied in high volume sprays, a wide range of droplet sizes are used. But with the current trend to reduce the volume of spray applied the droplet size assumes some importance (Matthews, 1975*a*). For example, for the control of flying insects, aerosols with extremely small droplets are used because they will persist in the air for an appreciable period and be picked up by the flying insect. But for spraying relatively immobile insects, a coarser spray with larger droplets must be used. Sprays (Anonymous, 1974; Matthews, 1975*b*) may be classified according to droplet size as follows:

Droplet size classification	Volume median diameter (vmd) of droplets
fog	5–15 μm
aerosols	16–50 μm
mists	51–100 μm
fine sprays	101–200 μm
medium sprays	201–400 μm
coarse sprays	>400 μm

A graticule for the determination of spray droplet size is described by Matthews (1975*b*).

Matthews (1973) classifies the nozzle, according to the form of energy employed to break up bulk liquid into droplets and to project the resultant spray droplets to the required target. The five categories of energy used in operating sprayers and mentioned by these authors are as follows:

(1) Hydraulic energy.
(2) Gaseous energy.
(3) Centrifugal energy.
(4) Kinetic energy.
(5) Thermal energy.
(6) Electrodynamic energy (since 1979).

Sprayers employing kinetic energy are now unavailable commercially. Thermal energy is used for fogging in warehouses. Sprayers based on the other forms of energy are extensively used in the tropics. The equipment used in Africa may be conveniently discussed as follows:

(1) *Manually-operated equipment*

Hand pumped sprayers and pre-pressurized sprayers are used in spraying crops. Simple syringes or hand sprayers, while useful in the control of medical vectors are now rarely used in crop protection in Africa (Matthews, 1981).

Knapsack sprayers (Fig. 17) are used throughout Africa and are especially successful in treating small acreages. These sprayers work on the principle of hydraulic energy. Here the pressure may be produced by a pump that converts mechanical energy to hydraulic energy. Knapsack sprayers usually have a 10–15 litre container with a hand-operated pump attached. Some sprayers have a small piston or diaphragm pump actuated by a handle which the operator pumps continuously to maintain pressure. The spray liquid passes from the container into a compression cylinder so that the spray is continuous and even. Knapsack sprayers are fitted with straps and can be carried comfortably on the back. The sprayer may be fitted with various types of hydraulic nozzles through which the spray liquid is forced under pressure. Further, they are relatively cheap and simple to use and maintain. However, their chief disadvantage is that both hands are required for operation and their use over long periods is therefore tiring. Techniques for testing the durability of lever-operated knapsack sprayers are discussed by Matthews *et al.* (1969). A boom with several nozzles can be used on lever operated sprayers.

Pressurized sprayers do not have to be pumped during operation. The spray is placed in the container which is then pressurized using a hand pump or compressed air. When the hand control is opened, the pressure forces the spray through the nozzle. The main advantage of pressurized sprayers is that as the operator is not pumping while spraying he can direct the spray more carefully. While selecting the equipment care should be taken to ensure that the materials will withstand both the pressure and the fatigue of repeated pressurization and depressurization. A pressure regulating valve should be used to avoid rapid drop in pressure while spraying.

(2) *Power-operated sprayers* (*Motorized knapsack sprayers*)
In these machines gaseous energy is used to break up bulk liquid into drops and project the resultant spray. Several methods are available to provide gaseous energy but air is the gas most frequently used. Power-operated knapsack sprayers (Fig. 19) can be used for production of mist which is so important in situations such as cocoa farms in Ghana where trees grow to heights exceeding 4 m. Hence motorized knapsack mistblowers must be used to project insecticide spray into the top of the canopy in order to apply a toxic dose of chemical to the feeding areas of the mirids (Clayphon, 1971, 1979). In this type of sprayer, the liquid is fed under pressure through a restrictor into a high velocity air stream provided by a fan driven by a two-stroke internal combustion engine. The air-blast shears the liquid and

Figure 19 Motorized knapsack sprayer delivering insecticide (Courtesy CIBA–GEIGY Ltd, Basel).

directs the droplets towards the target. Motorized knapsack mistblowers are extensively used in controlling pests of tree crops in the tropics. There were more than 200 000 of these in 1978 in Ghana alone (Clayphon, 1979). The mistblowers can be easily modified for ULV applications. However, in view of the extreme fineness of the spray produced by this method, the operator must be sure that the machine is actually spraying during operation. In some cases the addition of a colouring agent to the insecticide may solve the problem. For an evaluation of lever-operated knapsack and motorized knapsack sprayers, reference should be made to the work of Sutherland (1979).

(3) *Tractor-mounted sprayers*

These are used for spraying field crops. A large tank or container of up to 500 litres is mounted at the rear of the tractor which provides power for the pump. Larger tanks are usually carried on trailers. The tank is connected to a horizontal pipe with outlet nozzles situated at intervals along its length. The equipment should have sufficient ground clearance to pass over the crops without causing appreciable damage. With such equipment large acreages can be easily treated. Tractor-mounted sprayers are really not very relevant for use in Africa though some are used with special booms on coffee in East Africa and a few on some large estates.

(4) *Aerial spraying*

Application of insecticides by aerial spraying gives good results if applied with the correct spraying equipment. Formulation, droplet size and meteorological conditions are important factors to be considered in this highly specialized branch of spraying. Aerial spraying should be considered only if there are 20 or more hectares of the land to be treated. Aerial spraying is most widely used on irrigated cotton such as in the Sudan Gezira. In some countries such as Malawi trials were carried out when individual small-scale farmers organized their cotton fields into larger blocks to take advantage of aerial spraying. Fields for this purpose should ideally be rectangular in shape with no obstacles such as telephone lines, power cables or trees obstructing the field or in the immediate vicinity of the field. Aerial spraying is important for speed and access to fields where water and canals interfere with equipment. For efficiency and cheapness of application, the fields should, however, be as close as possible to the airstrip so as to treat them all during one run. Highly useful practical information on aerial spraying is available in many publications put out by companies such as CIBA-GEIGY, Switzerland.

(5) *Ultra-low-volume spraying* (ULV) (Fig. 20)
A small volume of a pesticide can be sprayed effectively over a large area by breaking it up into extremely small droplets using centrifugal energy. Instead of conventional nozzles a spinning disc device is used. Insecticide is stored in a plastic bottle and is gravity fed through a calibrated orifice onto the atomizer by inverting the sprayer. At the rear of the atomizer head there is a small 12V DC motor which drives a spinning disc. The motor is powered by dry cell transistor type batteries, either contained in the handle ('ULVA') (Fig. 20a, 21a) or in a separate power pack slung from a shoulder strap (Turbair 'X'). The spinning disc has a series of fine teeth – about 14 per cm around the circumference. These act as point sources for release of streams of very fine, uniformly sized droplets (less than 120 μm in diameter) which are then carried horizontally by wind and downward by gravity.

The ULV formulations are usually solutions with a higher concentration of active ingredients than with high volume sprays, hence special care must be exercised in handling them. Note should be taken of the wind direction and speed so that as little spray as possible falls on the operator.

Mowlam (1973) states that the introduction of ultra-low volume hand spraying, using light-weight, battery operated sprayers, has brought cotton spraying within the reach of many small African farmers who live far from regular water supplies. According to this author 70 000 ha and 110 000 ha of cotton crop were sprayed by this method in Mozambique and Tanzania respectively in 1973. The concept of ULV spraying for insect control was pioneered in Africa by the Desert Locust Control Organization during the early fifties (Sayer, 1959). During the last five years spraying with ULV hand machines has become increasingly popular for the control of cotton pests in Ghana where experience has shown that one set of 8 batteries is sufficient to spray 30 hectares. The ULV applicator with its one-litre insecticide container weighs less than 3 kg, and is capable of treating roughly two hectares per hour. In ULV applications there is no troublesome transporting of water; the hard task of pumping the knapsack sprayer under the hot tropical sun as well as its being carried on the back is also eliminated. In many ways waterless spraying is of immense potential benefit to agriculture in arid and semi-arid regions. As

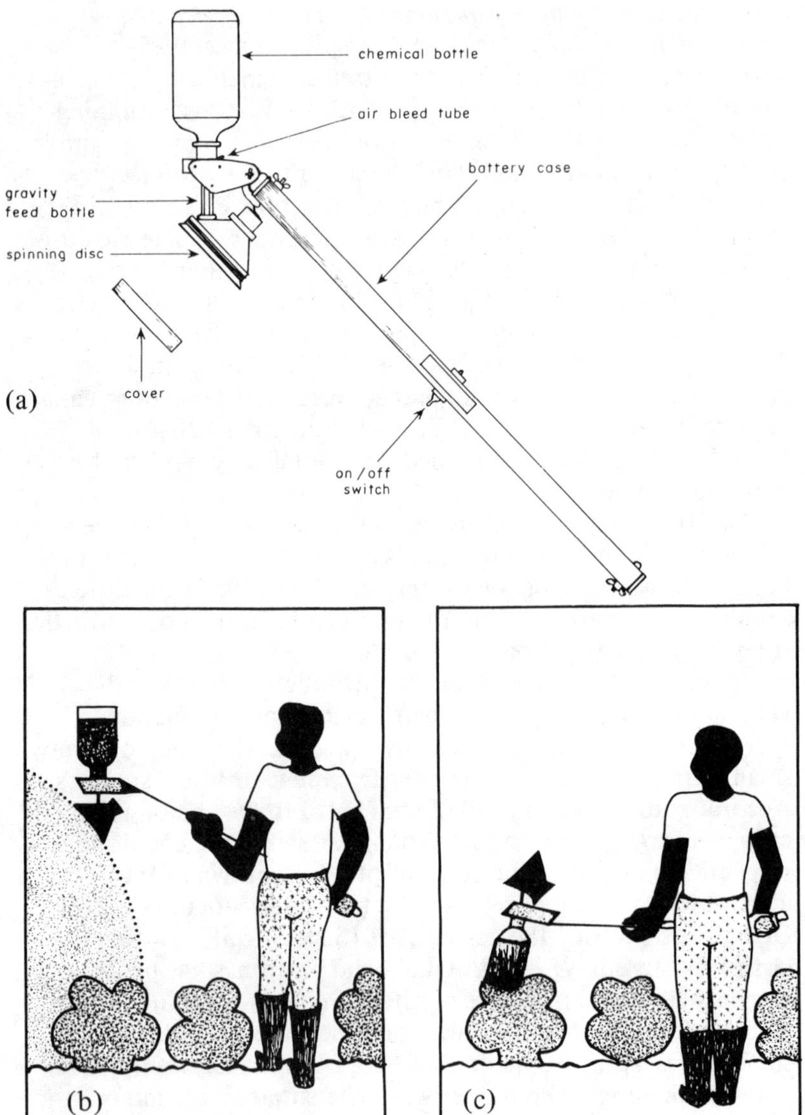

Figure 20 (a) Parts of a Micron 'ULVA' sprayer (after Matthews, 1976).
(b) Upward (spraying) position of ULVA sprayer: flow of liquid through the gravity feed tube.
(c) Downward (idle) position of ULVA sprayer at the end of a spray path (after Anonymous, 1977, courtesy CIBA–GEIGY Ltd, Basel).

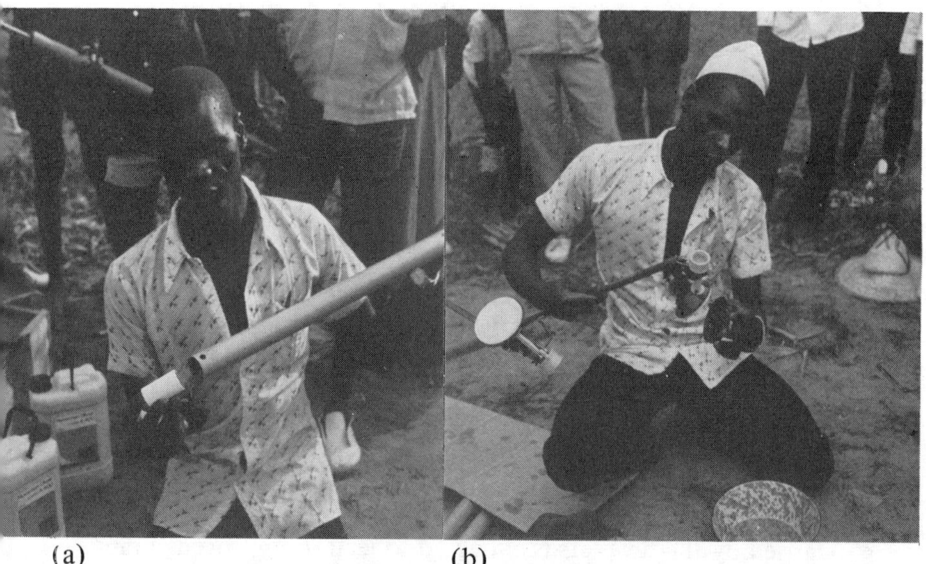

(a) (b)

Figure 21 (a) Fitting dry cell batteries in the handle of a ULVA–spinning disc sprayer (courtesy, CIBA–GEIGY Ltd, Basel).
(b) Farmer fitting the spinning disc of the ULVA sprayer (courtesy, CIBA–GEIGY Ltd, Basel).

noted by Matthews (1973), under tropical conditions less volatile formulations are essential as water evaporates rapidly.

In Malawi and the Gambia, WULV-water based ULV spraying using wettable powders has been used. Actually the cost of ULV spraying is reduced by the use of cheaper, finely micronized, wettable powder formulations, suspended in water (Matthews, 1976, 1981). With this technique the volume of liquid to be sprayed is increased to 10–12 1 ha^{-1} or more, and a narrow swath (1–2 m) is required to reduce the effect of evaporation, on droplet size. The actual usable swath width or spray path intervals depend on crop, growth stage, target site of the pest and the prevailing wind velocity. For treating pests exposed on tops of plants wide swaths are used. On pests feeding on the lower part of the plant narrow swaths are used. In Central Africa, the swath width is reduced from 6, to 4, and finally to 2 m swaths as the cotton plant grows.

Care and maintenance of hydraulic, motorized and ULV spraying equipment referred above is fully discussed by Clayphon & Matthews (1973), Bindra & Singh (1977) and

Matthews (1976, 1979). In addition there is a wealth of practical information on the principles, use and maintenance of spraying equipment by several international companies such as CIBA-GEIGY (Anonymous, 1977), ICI etc.

(6) *Electrostatic spraying*
Simple spraying machines which produce electrostatically charged droplets are currently in the process of development. The Electrodyn sprayer, an ICI development, not only produces electrostatically charged droplets, but utilizes electrostatic energy to achieve atomization thereby completely dispensing with all moving parts (Coffee, 1981). The machine is powered by torch batteries which in view of very low power consumption may give up to 50–60 hours use. The droplet spectrum is said to be very narrow and can be varied from 40–200 μm vmd to suit particular targets. The electrical charge carried by the droplets is expected to lead to improved spray distribution, particularly with regard to underleaf cover, and drift, wastage and operator contamination would be significantly reduced. Electrostatic spraying could cut pesticide use by as much as two-thirds and sharply reduce the amount of pesticide left in the atmosphere, or collected by run off in rivers and lakes. The application of the principles of electrostatics to spraying would seem to open up exciting new possibilities for the safer and more efficient use of pesticides.

11.2.2 Dusters and granule applicators

Dusting equipment As stated earlier, dusts have the advantage over water-based sprays in situations where water is scarce. Dusting machines basically consist of a container to hold dust which must be fed at a constant rate into the air stream produced by a manual or power-operated fan. Some means of agitation is usually incorporated in the container to prevent formation of lumps and ensure continuous flow during operation. Manually operated dusters are widely used in public health operations. Thus for treating rat burrows for flea control, crank- or hand-rotary type duster is often used. In the "bellow" duster, the air is supplied by a bellows instead of a rotary fan. Power-driven dusters can now be modified to apply granules which as noted by Anonymous (1974), can be projected into difficult terrain, e.g. under thick vegetation where dust or liquid formulations are inefficient.

11.3 Importance of correct application

Correct application of insecticides involves correct timing of spray, effective coverage of target and correct dosage. The timing of spray should be determined by a careful study of the life cycle of the pest. For example, it may be advisable to spray during early larval stages of the pest before it is capable of causing extensive damage. The spray timing will obviously need to be worked out for each pest situation. Correct timing not only provides good control but may also reduce the number of applications required for the purpose. This is especially evident in integrated control programmes (see Chapter 13).

The extent of control of the pest tends to be governed by the target coverage from the insecticide. Generally, excellent coverage provides excellent control, mediocre coverage leads to mediocre control and bad coverage results in bad control. For optimal efficacy of the insecticide application, the factors to be borne in mind include swath width, speed of application, spray volume/droplet size and flow rate. The proper calibration of the application equipment is an important necessity in getting correct dosage to the target. Many studies have shown that the volume of flow and the spray pattern both change as nozzles wear (Anonymous, 1979 *a*). The wear is highest when wettable powders are being applied. Therefore, the sprayer should be calibrated when applying such materials to ensure that the correct dosage of insecticide is being directed at the targets.

11.3.1 Preventive maintenance of equipment

No application equipment, however good, will last very long if preventive maintenance measures are not adopted. First of all the recommendations of equipment and chemical manufacturers should be closely observed. Contact time between chemical and equipment should be reduced to the barest essential. Unnecessary exposure of the equipment to heat and sunlight should be avoided. Application equipment should be inspected regularly as certain parts, especially those of sprayers, are liable to wear and tear. Components such as pistons, brackets, gaskets, valves, hoses, etc. should be carefully examined and spares of these should be kept to hand for replacement when necessary. As well as these, special attention should be paid to the following aspects of maintenance:

Cleaning At the end of each day's operations, spraying equipment should be emptied and cleaned thoroughly inside as well as outside. Nozzles and filters should be removed and cleaned with a brush and recommended cleaner. Spray lances should be disconnected from sprayers and the piston tube of compression sprayers removed from the spray tank. Metal parts should be coated with oil prior to storage in order to prevent rusting and consequent clogging of nozzles.

Storage Equipment should be stored in a dry shaded place. Various parts of the equipment should be kept disconnected as gentle ventilation tends to assist in the hardening of their interior surfaces.

For a comprehensive treatment of pesticide formulations, application equipment and its maintainance and methods of spray, reference should be made to the recent book by Matthews (1979). Useful information is also provided by Anonymous (1974) and Bindra & Singh (1977). It should perhaps be mentioned that in the economic realities of largely subsistence farming in Africa, the range of equipment available for pesticide application is severely limited. After preliminary advice and usually some trials under field conditions, a government generally imports in bulk a particular brand of recommended machinery as well as insecticide for sale, usually at subsidized rates, to the farmers. The wisdom of subsidy has been a matter of debate as if the chemical is too cheap, it may be wasted. Due to import and foreign currency restrictions, the chances of an individual or group of individuals importing the application equipment of their choice are usually rather remote. Matthews (1981) believes that the recent developments in electrostatic sprayers using very small amounts of insecticides may help to overcome the financial constraint to adequate crop protection for the peasant farmer.

11.4 Safety precautions for pesticide application

Handling pesticides poses one of the biggest threats to occupational safety in agriculture. The farmer not only needs to understand the danger to himself of using chemicals but also much wider issues of drift, disposal and storage. An American educational organization, the Association for Vocational Instructional Materials, recognizing this, has published (Anonymous, 1976) a practical manual on pesticide management with the aim of training agricultural workers in the 'effective, efficient and safe' application of insecticides. Matthews & Clayphon (1973) have produced a code of practice to be followed when using pesticides in the tropics. The details (with some modifications) are as follows:

(*a*) *Before applying pesticides – general instructions*

(1) Know the pest, and how much damage is really being done by it.
(2) Use pesticides only when really needed.
(3) Seek advice on the proper method of control.
(4) Use only the recommended pesticide for the problem. If several pesticides are recommended, choose the least toxic to mammals and if possible the least persistent. The use of most toxic pesticides will require too much protective clothing which will be highly uncomfortable under African conditions.
(5) READ THE LABEL including the small print.
(6) Make sure the appropriate protective clothing is available and is used, and all concerned with the application also understand the recommendations. For example, wear a long sleeved shirt, long trousers and a broad-brimmed rubber or plastic hat (NOT a cap). Use a face mask (piece of cloth, handkerchief, etc.) to prevent excessive inhalation of spray mist. Use unlined rubber boots and keep trouser-legs outside them to avoid draining of pesticides inside the boots. Wear unlined rubber gloves (NOT cotton or leather gloves) and wash them with water and detergent before removing. Do not wear contaminated clothes after spraying.
(7) Commercial operators using large quantities of

organophosphate pesticides, should visit their doctor and have a blood test, and have repeat checks during the season.

(8) Check application equipment for leaks, calibrate with water (for conventional sprayers other than ULV machines) and ensure it is in proper working order.

(9) Check that plenty of water is available with soap and towel and that a change of clean clothing is available.

(10) Warn neighbours of your spray programme, especially if they have apiaries.

(11) Check that pesticides on the farm are in a dry locked store. Avoid inhaling pesticide mists or dusts, especially in confined spaces such as the pesticide store.

(12) Take only sufficient pesticide for the day's application from the store to the site of application. DO NOT transfer pesticides into other containers, especially beer and soft drink bottles.

(*b*) *While mixing pesticides and during application*

(1) Wear appropriate protective clothing. If it is contaminated, remove and replace with clean clothing.

(2) Never work alone when handling the most toxic pesticides.

(3) Never allow children or other unauthorized persons near the mixing.

(4) Re-check the instructions on the label.

(5) Avoid contamination of the skin, especially the eyes and mouth. Liquid formulations should be poured carefully to avoid splashing. Avoid powder formulations 'puffing up' into the face. If contaminated with the concentrate, wash immediately.

(6) Never eat, drink or smoke when mixing or applying pesticides. To avoid contamination danger, do not carry cigarettes or edibles in your pocket while handling or spraying insecticides.

(7) Always have plenty of water available for washing.

(8) Always stand upwind when mixing.

(9) Make sure pesticides are mixed in the correct quantities.

(10) Avoid inhalation of chemical, dust or fumes.

(11) Start spraying near the downwind edge of the field and proceed upwind so that operators move into unsprayed areas.
(12) NEVER blow out clogged nozzles or hoses with your mouth.
(13) Avoid spray drift – DO NOT spray if wind conditions cause drift, as birds, bees and wildlife may be endangered. Never spray if the wind is blowing towards grazing livestock or regularly used pastures.
(14) NEVER leave pesticides unattended in the field.
(15) Provide proper supervision to those assisting with pesticide application, and have adequate rest periods.
(16) When blood tests are being conducted, do not work with pesticides if your cholinesterase level is below normal.

(*c*) *After application*

(1) RETURN unused pesticide to the store.
(2) Safely dispose of all empty containers. If it is difficult to bury empty containers after each day's spraying operations, they should be kept in the pesticides store until a convenient number are ready for disposal. IT IS ABSOLUTELY IMPOSSIBLE to clean out a container sufficiently well to make it safe for use for storage of food, water or as a cooking utensil. If any containers are burnt, NEVER stand in the smoke.
(3) NEVER leave pesticides in application equipment. Clean equipment and return to store. Wear protective clothing when cleaning spray equipment.
(4) Remove and clean protective clothing.
(5) Wash well and put on clean clothing. Where there is a considerable amount of spraying the operators should be provided with a shower room. It is essential to wash well with water and soap before drinking, eating and smoking.
(6) Keep a record of the use of pesticides. Note the formulation, concentration and duration etc. of each type of chemical sprayed in a log book. These records may help in detection of the origin of equipment deterioration.

(7) Do not allow other persons to enter the treated area for the required period if restrictions apply to the pesticide used.

(*d*) *Salient features of ARTIFICIAL RESPIRATION for patients affected by pesticides*

If the patient is not breathing respiratory resuscitation is of extreme urgency. Use the mouth to mouth or nose method (see below) and continue until the casualty is breathing satisfactorily or until a doctor tells you to stop.

Extreme pallor, widely dilated pupils of the eyes and failure to respond to the first few inflations of the lungs are evidence of circulatory failure. The heart has stopped and external cardiac massage must be applied as well as resuscitation.

(1) If possible lay patient on back with head a little higher than feet.
(2) Tilt the patient's head back and lift the jaw to move tongue away from the back of the throat, and if necessary clear the mouth of any obstruction.
(3) Make sure the face, especially around the mouth and nose, has been washed to remove any pesticide contamination.
(4) Take a deep breath and either close his mouth and blow firmly but gently into his nose;
or if the patient has not swallowed any pesticide pinch his nose and blow firmly but gently into his mouth. As you do this, the chest will rise. Turn your head away and take another breath watching for the chest to fall. Repeat with four quick breaths and then continue with one breath every five seconds until the patient starts to breathe on his own.

It is important to have a proper seal over his nose or mouth. If the patient vomits, turn patient on his side, clear his mouth and throat, then continue resuscitation as above. If air enters the stomach, as shown by a swelling of the upper part of the stomach, turn the patient away from you and apply gentle pressure to push the air out. This may cause the patient to vomit.

A manual on recognition and management of pesticide poisonings is by Morgan (1976). Although the text is aimed mainly at physicians treating patients, the information is also useful to investigators who may wish to confirm pesticide poisonings. First aid for pesticide poisoning is discussed by Anonymous (1979*b*). A practical guide on handling pesticides safely has recently been published by CIBA-GEIGY Ltd (Anonymous, 1981) while Matthews' (1979) book also has a detailed chapter on safety precautions.

11.5 Literature cited

Anonymous (1974). *Equipment for vector control.* World Health Organization, Geneva. 179 pp.

Anonymous (1975). Pesticide application equipment. *PANS*, **21(4)**, 436–49.

Anonymous (1976). *Applying Pesticides: management, application, safety.* American Association for Vocational Instructional Materials, Athens, Georgia, U.S.A. 96 pp.

Anonymous (1977). *Recommendations for Application – spinning disc sprayers for insecticides.* CIBA-GEIGY Ltd, Basel. 23 pp.

Anonymous (1979*a*). *Calibration of Ground Sprayers*. AG.8.11/HZ. CIBA-GEIGY Ltd, Basel. 71 pp.

Anonymous (1979*b*). First aid for pesticide poisoning. *World Farming*, **21(1)**, 20–3.

Anonymous (1981). *Handling Pesticides Safely – protective clothing and equipment*. AG11/H. Pfalzer, CIBA-GEIGY Ltd, Basel. 11 pp. + illustrations.

Bals, E. J. (1973). Some observations on the basic principles involved in ultra-low-volume spray applications. *PANS*, **19(2)**, 193–200.

Bindra, O. S. & Singh, H. (1977). *Pesticide Application Equipment*. Oxford and I.B.H. Publishers Co., New Delhi. 464 pp.

Clayphon, J. E. (1971). Comparison trials of various motorised knapsack mistblowers at the Cocoa Research Institute of Ghana. *PANS*, **17(2)**, 209–25.

Clayphon, J. E. (1979). The evaluation of a motorized knapsack mistblower. *PANS*, **25(4)**, 440–3.

Clayphon J. E. & Matthews, G. A. (1973). Care and maintenance of spraying equipment in the tropics. *PANS,* **19(1)**, 13–23.

Coffee, R. (1981). Electrodynamic Crop Spraying. *Outlook on Agriculture,* March 1981, 350–6.

Deutsch, A. E. (Ed.) (1976). *A World-wide Categorized Partial Listing for Manufacturers of Pesticide Application Equipment.* International Plant Protection Center, Oregon. 56 pp.

Free, J. B., Needham, P. H., Raccy, P. A. & Stevenson, J. H. (1967). The effect on honeybee mortality of applying insecticides as sprays or granules to flowering field beans. *J. Sci. Fd. Agric.,* **18**, 133–38.

Furmidge, C. G. L. (1972). General principles governing the behaviour of granular formulations. *Pestic. Sci.,* **3**, 745–51.

Furmidge, C. G. L. & Shenton, T. (1973). Formulation: some factors affecting pesticide performance. *SPAN,* **16(2)**, 82–4.

Furmidge, C. G. L., Hill, A. C. & Osgerby, J. M. (1966). Physico-chemical aspects of the availability of pesticides in the soil. 1 – Leaching of pesticides from granular formulations. *J. Sci. Fd. Agric.,* **17**, 518–25.

Hartley, G. S. & Graham-Bryce, I. J. (1980). *Physical Principles of Pesticide Behaviour: the dynamics of applied pesticide in the local environment in relation to biological response.* Academic Press, London, 2 vols. 1024 pp.

Marrs, G. J. & Middleton, M. R. (1973). The formulation of pesticides for convenience and safety. *Outlook on Agriculture,* **7(5)**, 231–5.

Matthews, G. A. (1973). Nozzles for pesticide application in the tropics. *PANS,* **19**, 583–600.

Matthews, G. A. (1975*a*). Determination of droplet size. *PANS,* **21(2)**, 213–25.

Matthews, G. A. (1975*b*). A graticule for classification of spray droplets. *PANS,* **21(3)**, 343–4.

Matthews, G. A. (1976). New spraying techniques for field crops. *World Crops,* **28**, 106–10.

Matthews, G. A. (1979). *Pesticide Application Methods*. Longman, Harlow, 334 pp.

Matthews, G. A. (1981). Developments in pesticide application for the small-scale farmer in the tropics. *Outlook on Agriculture,* **10(7)**, 345–9.

Matthews, G. A. & Clayphon, J. E. (1973). Safety precautions for pesticide application in the tropics. *PANS,* **19(1)**, 1–12.

Matthews, G. A., Higgins, A. E. H. & Thornhill, E. W. (1969). Suggested techniques for testing the durability of lever-operated knapsack sprayers. *Cotton Growing Review,* **46**, 143–8.

Morgan, D. P. (1976). *Recognition and Management of Pesticide Poisonings*. U.S. Environmental Protection Agency. Office of Pesticides Programs, Washington, D.C. 20460. 49 pp.

Mowlam, M. D. (1973). Spraying cotton with ULV hand machines. *SPAN,* **16(3)**, 127–8.

Page, A. B. P. & Lubatti, O. F. (1963). Fumigation of insects. *Ann. Rev. Ent.,* **8**, 239–64.

Phillips, F. T. (1968). Microencapsulation: a method for increasing specificity and controlling persistence of pesticides. *PANS,* **14(3)**, 407–10.

Sayer, H. J. (1959). An ultra-low-volume spraying technique for the control of the desert locust, *Schistocerca gregaria* (Forsk.). *Bull. ent.*

Res., **50**, 371–86.

Sutherland, J. A. (1979). The evaluation of lever-operated knapsack and motorised knapsack sprayers. *PANS*, **25(3)**, 332–64.

Tunstall, J. P. & Matthews, G. A. (1972). Insect pests of cotton in the old world and their control, pp. 46–59. In *Cotton.* CIBA-GEIGY Agrochemicals. Technical Monograph No. 3 Basel, Switzerland. 80 pp.

Van Valkenburg, W. (Ed.) (1973). *Pesticide Formulations*. Marcel Dekker Inc., New York. 481 pp.

Walker, J. O. (Ed.) (1980). *Spraying systems for the 1980s*. Monograph No. 24. Proceedings of a symposium held at Royal Holloway College, Surrey, England. March 26–27th, 1980. BCPC Publications, 144–50 London Road, Croydon CRO 2TB. 319 pp.

12 Attractants, Repellents, Antifeedants and Hormones

Relatively simple chemical compounds mediate a number of vital life processes of insects by influencing their behaviour in searching for mates, food, shelter and sites for deposition of eggs. Similarly a number of metabolic process are governed by the release of hormonal secretions. Under normal conditions these chemicals are released at particular times and elicit specific behavioural and physiological activities. Modern methods of control seek to utilize such chemicals or their synthetic analogues in the manipulation of pest populations by modifying insect behaviour or disrupting normal physiological processes of insects. Such chemicals may be attractants, repellents, antifeedants, hormones or their analogues.

12.1 Attractants

Some insects possess a highly specialized sense of smell and are able to follow odours and trails to sources of food, oviposition sites and to mates. Attractants should provide directional clues that a flying insect can use at some distance in orienting towards the odour source. But this term has often been misused (Shorey, 1977*a*). For example, the term sex attractant is used when pheromone would be more accurate. Pheromones have been considered primarily to attract the opposite sex, but this is not always the case. Sex pheromone is not synonymous with sex attractant (Dethier *et al.*, 1960; Kennedy, 1972) as the former may serve as a locomotory stimulant, an attractant, an arrestant, and/or sexual stimulant (Shorey, 1977*a*). Currently, considerable research is in progress on the use of behaviour-modifying chemicals (see the books edited by Shorey & McKelvey (1977) and Rockstein (1978) for much of the latest information). The realization that there is more to chemical communication than sexual attraction has necessitated a more detailed classification of the substances involved. When individuals of different species communicate chemically, the signal is an allomone if the sender benefits. An example is those defensive secretions produced by many insects that are poisonous or repugnant to attackers. If the receiver benefits, however, the signal is a kairomone, as in the case of certain predators of bark beetles which aggregate in response to secretions from their prey. Only pheromones and food lures which have some application in the control of agricultural pests are discussed below.

12.1.1 Pheromones

A pheromone is defined as a chemical or a mixture of chemicals that is released to the exterior by an organism and that causes one or more specific reactions in a receiving organism of the same species (Shorey, 1977*b*). They are effective in minute quantities and act as chemical messengers between individuals. Pheromones are essential to the survival of many species, being the key to well-coordinated social behaviour patterns. Sexual behaviour, sexual maturation, trail following, alarm and individual or colony recognition are among the vital functions mediated by pheromones (see Law & Regnier, 1971; Shorey, 1973; Borden, 1977; Fletcher, 1977; and Tamaki, 1977, for some reviews of the subject).

As noted by Law & Regnier (1971) most pheromones are simple compounds of relatively low molecular weight and many are derivatives of fatty acids or terpenes. Pheromones may be multi-component (composed of several chemicals) or single component chemicals. While recently there has been some tendency to over-emphasize the multi-component aspect of pheromone systems, several highly effective 'single component' pheromones (e.g. those of gypsy moth and cabbage looper), are in use. Single component pheromones, pheromone mimics and even individual compounds of multi-component pheromones are valuable or potentially valuable in pest management systems.

Sex pheromones have been found in over 200 species of insects, especially among the Lepidoptera. Law & Regnier (1971), Tamaki (1977), and Anonymous (1979) among several recent works, document the chemical structures of many well-characterized pheromones. In the developing countries, largely as a result of work by the Scientific Units of the British Overseas Development Administration, female sex pheromones of twelve moth species have been identified and synthesized (Campion & Nesbitt, 1981, see Table 10). The possibilities for the usage of pheromones for pest control are still developing and their field applications may be divided into three categories:

1) *Stimulation of specific behaviour patterns* An example is where the insects are attracted to traps baited with sex pheromone and are killed or sterilized. Trammel *et al.* (1974) did mass sex-pheromone trapping of males for control of red-banded leaf roller (*Argyrotaenia velutinana*) in apple orchards. Pheromone-baited sticky traps were deployed, generally one per tree, prior to moth emergence in the spring. These traps maintained commercially acceptable and successful control of the pest. Mass trapping, i.e. large scale capture of pests in traps baited with pheromones forms the basis of several programmes throughout the world. The booklet by Roelofs (1979) provides a practical guide to the efficacy of mass trapping as a means of reducing crop damage in North America and includes an up to date bibliography.

Shorey (1977*b*) in reviewing the subject emphasized the care required in the design of an effective trap. The variables to be considered include the size, colour, and shape of the trap, the size of trap orifice for entrance of insects (see also Steck & Bailey, 1978) as well as the height above the ground at which the trap is set (see Shorey (1977*b*) for references). Furthermore, the concentration of the pheromone and the mechanism whereby the pheromone is slowly released may also be of importance (Campion *et al.*, 1981).

Campion & Nesbitt (1981), in discussing the use of mass trapping, state that where cheap labour is available, deployment of limited number of traps per unit area to catch the pest may be more convenient and economical compared with insecticide use. These authors report progress, in Egypt, on the control of Egyptian cotton leaf worm, *Spodoptera littoralis*, by mass trapping. This pest has been reported to be successfully controlled in Israel with pheromone-baited traps (Teich *et al.*, 1979).

2) *Survey and monitoring* The use of pheromones for survey and monitoring is now quite common especially in the United States (see Shorey, 1977*b* for references). Pheromone-baited traps help to monitor variations in the size of insect populations so as to provide information for the use of insecticidal control methods. These traps have the added advantage of attracting only the target species thus obviating the time consuming and laborious process of sifting the trap contents. They require no electricity to operate and can be used in remote areas.

Medlure, an empirical synthetic lure of Mediterranean fruit-fly (*Ceratitis capitata*), is used in the United States to detect infestation at an early stage. Some 20 000 traps are, for example, maintained in Florida. The first success was in 1962 when medlure detected incipient populations of Mediterranean fruit-fly and these were easily eliminated, perhaps saving the 9 million dollars used in the control of outbreaks in 1956. Thousands of traps using dispalure, the sex pheromone of the gypsy moth are used to monitor expansion of the pest across the United States of America.

In codling moth (*Cydia pomonella*) the pheromone is used to monitor orchards as a guide to spraying. In New York State orchards, pheromone traps monitor pest incidence of six species of caterpillars, providing weekly advisory data.

A large number of pheromone traps are now being used in monitoring the population of the armyworm in the outbreak areas in East Africa and thus help in providing an early warning system about incipient outbreaks of this highly destructive and migrant pest. Campion & Nesbitt (1981) provide details of a number of projects on monitoring now in progress in the developing world. These include the combining of the attractant pheromones of *D. castanea* with those of the pink bollworm (*Pectinophora gossypiella*) in order to develop a dual monitoring system in Shire valley cotton growing region of Malawi (Marks 1976, 1977). Close correspondence between catches of *S. exempta* in E. Africa in pheromone traps and light traps suggests that the economic thresholds originally based on light trap data could be readily adapted to the pheromone trap system (Campion & Nesbitt, 1981).

Survey and monitoring by pheromones is now an active field of research. Monitoring is already routine in many parts of Europe and North America, with commercially-produced traps,

Table 10 Pheromones identified, synthesized and field tested in developing countries (after Campion and Nesbitt, 1981, modified; reproduced by permission of the Controller, Her Majesty's Stationery Office.

Insects and crops attacked	Distribution	Pheromone components
Red bollworm (*Diparopsis castanea*) Cotton	South-eastern Africa	9, 11-dodecadienyl acetate (80% Z, 20%E), 11-dodecenyl acetate } attractant (E)-9-dodecenyl acetate, inhibitor Dodecanyl acetate, probably a pheromone precursor
Egyptian cotton leafworm (*Spodoptera littoralis*) Cotton, lucerne, potatoes, tomatoes, groundnuts	Africa, Mediterranean countries	(Z)-9, (E)-11-tetradecadienyl acetate, attractant (Z)-9-tetradecenyl acetate, inhibitor (E)-11-tetradecenyl acetate, Tetradecenyl acetate } probably precursors
African armyworm (*Spodoptera exempta*) Cereals, grasses	Africa, S.E. Asia, Philippines, New Guinea, Australia	(Z)-9-tetradecenyl acetate, (Z)-9, (E)-12-tetradecadienyl acetate } attractant
Striped stem borer (*Chilo suppressalis*) Rice	India, S.E. Asia, Indonesia, Japan	(Z)-11-hexadecenal, (Z)-13-octadecenal } attractant Synthetic 'pheromone mimics': (Z)-9-tetradecenyl formate, (Z)-11-hexadecenyl formate } inhibitors
Pink borer (*Sesamia inferens*) Rice, maize, sugar-cane	India, Malaysia, Indonesia, China, Japan	(Z)-11-hexadecenyl acetate, attractant

Table 10 (*continued*)

Pest / crop	Distribution	Compounds
Citrus flower moth, (*Prays citri*) Citrus	Mediterranean countries, India, Malaysia	(Z)-7-tetradecenal, attractant
Spotted stalk borer (*Chilo partellus*) Ceareals	India, East Africa	(Z)-11-hexadecenal, attactant (Z)-11-hexadecen-1-ol
Olive moth (*Prays loeae*) Olives	Mediterranean countries	(Z)-7-tetradecenal, attractant
Spiny bollworm (*Earias insulana*) Cotton	Africa, India, Mediterranean countries, Pakistan, S.E. Asia	(E,E)-10, 12-hexadecadienal, attractant
American bollworm (*Heliothis armigera*) Polyphagous, in particular cotton	Widespread, old world tropical and sub-tropical regions	(Z)-11-hexadecenal } attractants (Z)-9-hexadecenal } (Z)-11-hexadecenal-ol-inhibitor Hexadecenal } probably precursors Hexadecanol }
Sugar-cane borer (*Chilo sacchariphagus*) Sugar-cane	Java, Mauritius, Sumatra	(Z)-13-octadecenyl acetate } attractant (Z)-13-octadecen-l-01 }
Maize stalk borer (*Busseola fusca*) Maize and sorghum	Africa	(Z)-11-tetradecenyl acetate } (E)-11-tetradecenyl acetate } attractant (Z)-9-tetradecenyl acetate }

pheromones and dispensers being used to forecast the size of populations. On the basis of the numbers caught in the traps, farmers and their advisors decide how often to spray with pesticides and in many areas there has been a general reduction in the frequency of spraying. A commercial organization, Zoecon Corporation (California, U.S.A.), sells pheromone-monitoring kits for several species of insect pests. The obvious disadvantage of the use of highly selective chemicals such as pheromones is the high cost of development and low returns on investments. They are uneconomic to develop for pests of low-value crops.

3) *Confusion of specific behaviour patterns* (*communication disruption*) Here, in a habitat permeated with their own sex pheromone, it is anticipated that males will become confused and fail to find a large proportion of the females with which they would normally mate. This has been considered promising for the control of lepidopteran pests (Beroza 1976; Roelofs *et al*., 1976, Shorey, 1976, etc). While no spectacular or effective control of insect populations by this method has been reported, some encouraging reports have been made in the use of pheromones for pest control. Shorey *et al*. (1967) calculated that in the cabbage looper (*Trichoplusia ni*) it is sufficient to permeate the atmosphere with less than 0.5 mg ha^{-1} of female pheromone (*cis*-7-dodecenyl acetate) per night to completely prevent orientation of males to pheromone-emitting females.

Attempts to control pink bollworm (*Pectinophora gossypiella*) moths by confusing males have produced mixed results. For example, McLaughlin *et al*. (1972) failed to reduce field infestations by permeation of the air in small plots in cotton fields with hexalure (*cis*-7-hexadecenyl acetate). Hexalure is a male attractant but not the sex pheromone of *P. gossypiella*. The natural pheromone, gossyplure, was recently identified as a mixture of the *cis*,*cis* and *cis*,*trans* isomers of 7, 11-hexadecadienyl acetate. Shorey *et al*. (1974) continuously evaporated hexalure into the air of cotton fields throughout the growing season. The consequent disruption of pheromonal communication between males and females of the pink bollworm resulted in a reduction in larval boll infestation comparable to that provided by commercial insecticide applications. According to Flint *et al.* (1974) during early trapping trials, traps baited with 500 μg of gossyplure (1:1 mixture of isomers) captured about 70 times more adults of *P. gossypiella* than traps baited with 25 mg of hexalure. The authors believe that using suitable, baited gossyplure traps, control of this pest is feasible. Smith *et al.* (1978) on the other hand reported that, in laboratory experiments, adult *P. gossypiella* mated as frequently in a gossyplure-permeated atmosphere as in an untreated atmosphere and the male moth flight was at random in both the atmospheres. This behaviour was considered to be an effective mate acquisition strategy within a host patch but may thwart attempts to control the pink bollworm by chemical confusion with gossyplure in fields previously planted to cotton. These authors doubted that confusion with gossyplure could succeed as an effective control strategy. They reasoned that male mass trapping was obviously a more promising system for deployment of gossyplure in a pink bollworm control programme as captured males would at least be removed from the pest population.

According to Shorey (1977*b*), disruption may also be accomplished by the use of parapheromone (nonpheromone chemicals that elicit behaviour identical to that caused by natural pheromones) (Kaae *et al.*, 1973), or by antipheromones – non-pheromone chemicals that directly block or inhibit responsiveness of insects to their natural pheromones. Beevor & Campion (1979) discuss the field use of inhibitory components of lepidopteran sex pheromones and pheromone mimics to modify the behaviour of the male insect and prevent location of the female. Large scale mating disruption trials using kilogram quantities of inhibitory components of *S. littoralis* and *D. castanea* have been carried out in Crete and Malawi (see Campion & Nesbitt 1981, for references). The trials have provided valuable scientific information and led to the development of micro-encapsulated formulations which protect the pheromone under field conditions and release it continuously and uniformly over several weeks (Campion *et al.*, 1981).

12.1.2 Food lures

Food lures have been used with limited success but their use has seldom been on a large scale. They may serve to attract the insects, causing them to orientate to the food source. A major advantage of food lures is that both sexes of many species respond to them. The first attractants were essentially food lures based on combinations of natural plant products and flavouring essences. The latter may be token representatives of the nutritive components of suitable foods. The floral odours

Species	**Lure**	**Structure**
Mediterranean fruit-fly (*Ceratitis capitata*)	Trimedlure: t-butyl-4 (or 5)-chloro-2 methyl cyclohexane carboxylate	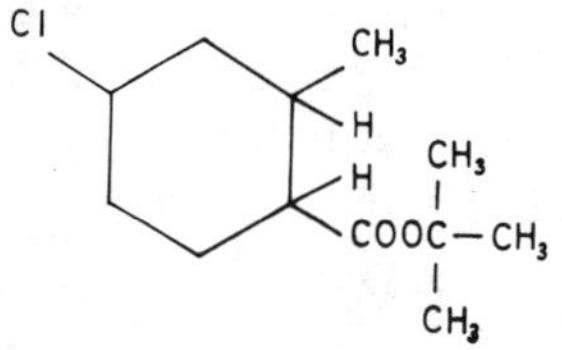 TRIMEDLURE
Fruit-flies (*Dacus* spp.) (some economic species do not respond to this attractant)	Cue-lure: 4-(p-acetoxyphenyl) -2-butanone	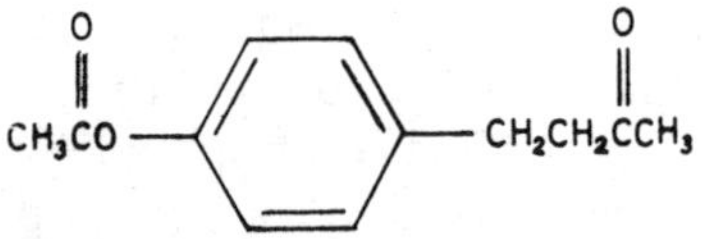 CUE - LURE
Oriental fruit-fly (*Dacus dorsalis*) and some other *Dacus* species (Drew, 1974)	Methyl eugenol: O-methyl eugenol	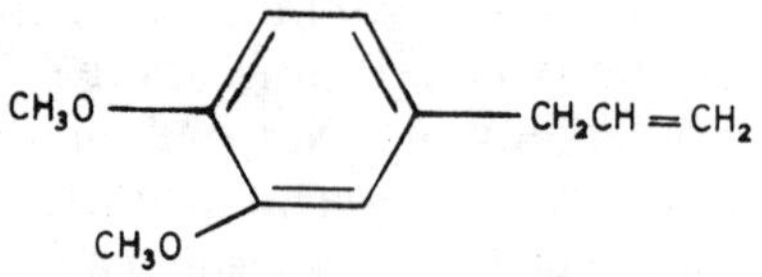 METHYL EUGENOL
Rhinoceros beetle (*Oryctes rhinoceros*)	Ethyl 3-isobutyl-2,2-dimethyl cyclopropane carboxylate	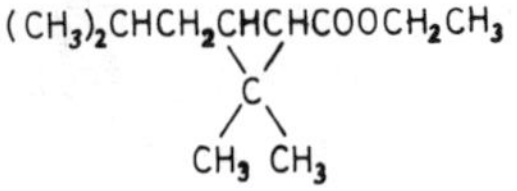 $(CH_3)_2CHCH_2CHCHCOOCH_2CH_3$ (ring C bearing CH_3 CH_3)
Yellowjackets (*Vespula* spp.) (MacDonald *et al.*, 1973)	Heptyl butyrate	

Figure 22 Food lures of some well known insects.

are usually a good attractant for nectar feeders whereas haematophagous insects tend to be attracted by carbon dioxide (see Knox & Hays, 1972 for some references on the subject) and lactic acid.

The most powerful attractants are chemicals which attract male insects and can be referred to as parapheromones. However, insects such as fruit-flies, feed vigorously on cue-lure and methyl eugenol when attracted and in this regard behave like a food lure (Drew *et al*., 1978). These authors recommend a mixture of 4 ml attractant and 1 ml 50% w/v concentrate of malathion or dichlorovos when charging a fruit-fly trap. They also state that the parapheromones attract flies over larger distances than do the food lures. Distances of up to 0.8 km have been suggested in the case of methyl eugenol. Jacobson (1965, 1966) and Beroza (1970, 1972) provide a general treatment of insect attractants while Madsen (1967 and later papers) deals with codling moth attractants. Food lures of male insects for some well known insects are shown in Fig. 22.

12.2 Repellents

Repellents are chemicals that prevent damage to plants, animals, or materials such as fabrics and timber by rendering them unattractive, unpalatable or offensive (Metcalf & Metcalf, 1975; Bottrell, 1979). Dethier *et al*. (1960) define a repellent as a chemical that causes animals to make oriented movements away from its source. Repellents have, as yet, proved to be of little value in the control of crop pests but a number of chemicals have been recommended for the control of livestock pests. These include synergized pyrethrins to protect cattle against biting flies. Compounds such as butoxypoly-propylene glycol, dipropylpyridine-2, 5-dicarboxylate, 3-chloropropyl octyl sulfoxide etc. are usually formulated with pyrethrins. A number of preparations that repel mosquitoes from the human body are also available (see Critchley, 1971 for some common insect repellents). Some of the newer repellents are said to look promising against a broad range of pests (Schreck, 1977). Packaging materials are frequently treated with chemical repellents on the outer surface to prevent or restrict insect penetration. Metcalf & Metcalf (1975) believe that the use of foliage repellents offers few advantages in pest control programmes. Not only do these chemicals require a very thorough coverage to prevent insect damage but present about the same degree of environmental hazards as conventional pesticides (Metcalf & Metcalf, 1975). Glass (1976), however, indicates great potential for use of effective repellents to protect plants from insect damage but no repellent has yet been found useful for protecting crops in the field or storage.

12.3 Antifeedants

Substances which interfere with the feeding activity of a pest on the treated plant offer yet another novel approach to insect pest control. Antifeedants are defined as substances which, when tasted, can induce cessation of feeding either temporarily or permanently, depending upon the potency (Nakanishi, 1977). Several authors have tried to distinguish between 'true antifeedants' and 'feeding deterrents' (see, for example, Higgins & Pedigo, 1979). A perusal of the literature shows that such terms as feeding repellent, rejectant, inhibitory chemical and feeding deterrent are often used synonymously with antifeedant. Generally these chemicals retard the feeding activities of pests and reduce damage caused by them by rendering treated plants unattractive and unpalatable. The insect often remains near the treated foliage containing the antifeedant and may die from starvation. Antifeedant compounds for insect control have been dealt with by Ascher & Nissum (1964), Wright (1963, 1967) and Chapman (1974) has provided a general review of the subject.

Properties of antifeedants (see Chapman, 1974 for references)
(1) Persistent
(2) Systemic, i.e. translocated to growing points of treated plants. Otherwise, the new plant growth developing after treatment will be selectively attacked.
(3) Without harmful effects on non-target organisms.

Examples of insect antifeedants Chapman (1974) notes that field trials on the practical use of antifeedant compounds were carried out between 1930 and 1940, mainly in the United States. Interest in these compounds then appears to have waned until the 1960s when the disadvantages of insecticides led to renewed interest in the antifeedants. Recent work has generally centred on the study of insect antifeedants in plants. Extracts of leaves or seed kernels of neem or Persian lilac when applied to plant foliage, or incorporated into a diet, have been reported to adversely affect the development of a number of insect species (see for example, Munakata, 1977; Nakanishi, 1977; Meisner *et al*., (1978, etc.). Meinwald *et al*., (1978) presented a summary of insect antifeedants from East African plants. Their work has introduced new antifeedants such as diterpenoids and suggests that many plants have evolved biosynthetic pathways for the production of a wide variety of highly functionalized terpenoids that are capable of providing defence against insect attack. Recently, Saxena *et al*. (1981*a*) reported a drastic reduction in feeding by the brown plant-hopper (*Nilaparvata lugens*) on rice plants treated with 'Neem', *Azadirachta indica*, seed oil (crude oil, extracted from decorticated neem seed). Longevity, fecundity and oviposition of the plant-hopper were significantly reduced on neem-oil-treated plants. Neem oil was also reported to have significantly deterred feeding by another rice pest, the rice leaf folder, *Cnaphalocrocis medinalis* (Saxena *et al*., 1981*b*). Azadirachtin, an active component of neem oil, has been found to be a powerful feeding deterrent to *Schistocerca gregaria*, as well as being a systemic growth disrupter (Meinwald *et al*., 1978) and is generally non-toxic to vertebrates. Saxena *et al*. (1981*b*) report that confinement of 5th instar larvae of the rice leaf folder on plants treated with at least 12% neem oil resulted in development of larval-pupal monstrosities which retained larval cuticle patches, head capsule, thoracic legs etc. Further, moths emerging from apparently normal-looking pupae often had poorly-developed or twisted wings. Despite the unique properties of neem seed oil in their effectiveness against a number of insect pests, little work is in progress in Africa in the use of this compound against crop pests. Chapman (1974) quite correctly states that antifeedants are no less costly to use than are insecticides, especially when repeated applications may be necessary. Further, identification and synthesis of antifeedants is hardly likely to be economic. But the neem tree is abundant in many African countries. The effectiveness of its oil even in a crude state against pests, safety to environment, low cost and ready availability suggests that its potential for pest control be fully explored. From the information available so far, it seems that antifeedants have an important role to play in the development of effective pest control strategies.

12.4 Insect hormones and their analogues

The earliest insecticides were well known poisons such as the inorganic compounds of lead arsenate, copper sulphate and even sulphur. To this first generation of insecticides one could add plant products such as rotenone and pyrethrum powders.

The second generation of insecticides is exemplified by DDT and related organochlorines, organophosphates and methyl carbamates. Their toxicity and stability represented insecticidal virtue and environmental vice. However, an increased realization of the hazards posed by their continued use prompted search for alternative methods of control.

A third generation of chemicals effective in insect control arose from basic studies of insect endocrinology and biochemistry. Insect hormones are internal secretions that regulate a wide range of physiological processes, including growth, development and maturation. The use of insect hormones as insecticides was advocated by Williams (1960) who called them the 'Third Generation Insecticides' (Williams, 1967). Generally insects important in the adult stages as disease vectors can be controlled in larval stages by hormonal chemicals. But for some insects the immature stages are the most damaging and these generally remain unaffected by hormone treatments. Hormones most likely to be useful as pesticides have been discussed by a number of workers, notably Ellis (1968) and Staal (1975, 1977). Biochemistry of insect hormones and insect growth regulators has been discussed by Riddiford & Truman (1978).

12.4.1 Moulting hormone (MH)

Moulting hormone, or ecdysone (Fig. 23), was characterized by Karlson *et al.* (1963) who found it to be a steroid having the same carbon skeleton as cholesterol. It has been synthesized and is available in quantity. However, this substance needs to be injected into the insect body in order to interfere with the

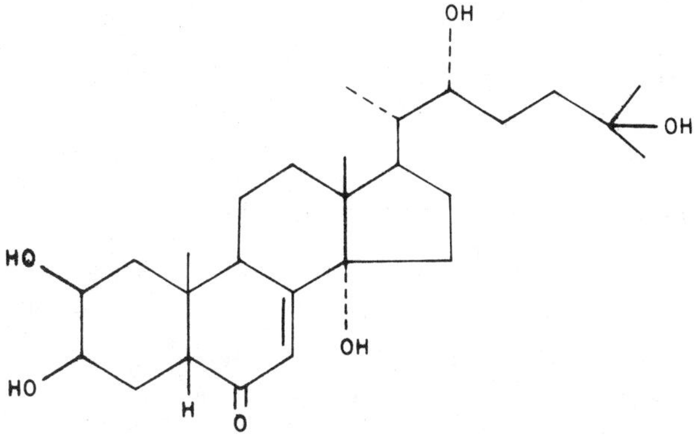

Figure 23 α-Ecdysone.

moulting process. Since ecdysone and its mimics are steroids and more hydrophilic than lipophilic, they cannot penetrate the waxy layer of the cuticle and so are ineffective when applied topically (Riddiford & Truman, 1978). Carlise & Ellis (1968) found that ecdysone was not absorbed into the body of the locust through the alimentary tract even when presented in baits. In others, at concentrations high enough to cause precocious moulting, ecdysone deterred feeding on the treated material (Riddiford & Truman, 1978).

12.4.2 Juvenile hormones (JH) and insect growth regulators

Juvenile hormones (JH) are secreted by the corpora allata of insects at specific intervals in the life cycle, and maintain the insect in an immature state until conditions favourable for maturation have been achieved (Schooley, 1977). These hormones are also known to be essential, in many insect species, for pheromone production by adult females and for stimulating egg maturation. These insect juvenile hormones (JH I, II, III) (Fig. 24) have been isolated and characterized from various species representing several insect orders (see, for example, Roller, *et al*., 1967, Schooley *et al*., 1976, and Trautmann *et al*., 1976). Recently, a new insect juvenile hormone (JHO) has been isolated from developing embryos of the tobacco hornworm moth, *Manduca sexta* (Bergot *et al*., 1980). The new hormone was found with juvenile hormone I and is a 1-carbon homologue of this substance.

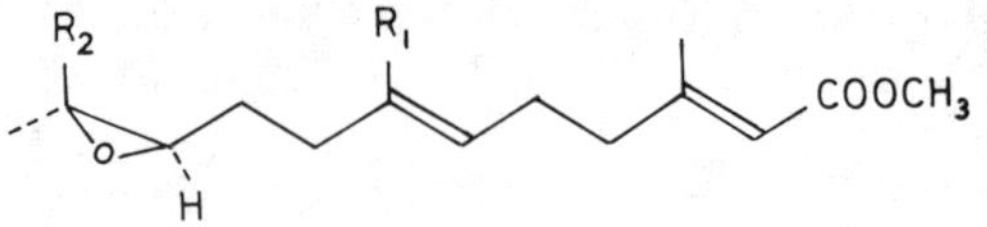

1 $R_1 = R_2 = -C_2H_3$ JH-I

2 $R_1 = -CH_3$, $R_2 = -C_2H_5$ JH-II

3 $R_1 = R_2 = -CH_3$ JH-III

Figure 24 Structure and nomenclature of 3 juvenile hormones.
(1) Methyl (2E, 6E, 10*cis*)-(10R, 11S)-10,11-epoxy-3, 11-dimethyl-7-ethyl-2,6-tridecadienoate (methyl 12, 14-dihomo juvenate, C_{18} Jh).
(2) Methyl (2E, 6E, 10*cis*)-(10R, 11S)-10, 11-epoxy-3,7, 11-tri-methyl-2, 6-tridecadienoate (methyl 12-homo juvenate, C_{17} JH).
(3) Methyl (2E, 6E)-(10R)-10, 11-epoxy-3, 7, 11-trimethyl-2, 6-dodecadienoate (methyl juvenate, C_{16} Jh).

Some plants such as balsam fir (*Abies balsamea*), used in the manufacture of paper, contain JH-like compounds called 'paper factor' which prevent insects such as *Pyrrhocoris apterous* from completing their development. Bowers *et al.* (1966) obtained an active compound, juvabione (methyl ester of todomatuic acid), from balsam fir wood distillate. Another compound, dehydrojuvabione (4(2,6-dimethyl-4-oxo-2-hexenyl)-1-cyclohexen-1-carboxylic acid methyl ester), was obtained by Černy *et al.* (1967) from the same wood. These 'paper factors' have been ineffective on hemipteran genera such as *Oncopeltus, Rhodnius, Oxycarenus* and various Orthoptera, Lepidoptera and Coleoptera. Even on *Pyrrhocoris* and *Dysdercus*, the 'paper factors' have no effect unless applied at about the time of the moult from the fourth to the last nymphal instar (Sláma & Williams, 1966*a*). Eggs laid by normal *Phyrrhocoris* adults fail to develop if placed in contact with active papers soon after laying (Sláma & Williams, 1966*b*). Černy *et al.* (1967) have been able to synthesize materials similar to dehydrojuvabione but a hundred times as potent. Little practical application has, however, been possible of discoveries such as these.

Nomenclature of the chemicals and mixtures with JH activity has given rise to considerable confusion. The terms JH analogue

(JHA), juvenoid, JH mimic, etc., all adequately describe this class of compounds. Staal (1975) uses the term juvenile hormone (JH) for natural hormones; insect growth regulators (IGR) when compounds are used with an intention to control pests, and on occasion JHA for less defined situations. IGR, JH, and JHA may thus be occasionally used for the same compound. The above scheme is followed in this book.

Control of medically-important Diptera A number of IGRs, e.g. RO-20–3600 (6,7-epoxy-3,7-dimethyl-1-(3,4-(methylenedioxy) -phenoxy)) -2-nonene) have been found to be quite effective in inhibiting growth or emergence of insects such as mosquitoes (*Aedes nigromaculis*) (Schaeffer & Wilder, 1972). According to Staal (1977), methoprene or Altosid® (isopropyl (2E, 4E)-11-methoxy-3,7,11-trimethyl-2,4-dodecadienoate) is, in fact, the most active larvicide ever developed. This compound has been approved by the Environmental Protection Agency in the United States for control of mosquitoes. Another compound, 1-(4-ethyl phenoxy)-6,7-epoxy-3,7-dimethyl-2-octene, has also shown promise as a mosquito larvicidal agent and has been accepted by WHO for mosquito control trials in several countries.

According to Wright *et al.* (1973), the stable fly, *Stomoxys calcitrans*, can be controlled by applying two juvenile hormone analogues as spray applications to larval breeding sites. Harris *et al.* (1973) reported inhibition of the development of the horn fly, *Haematobia irritans* and the stable fly, *Stomoxys calcitrans* in the faeces of cattle to which methoprene was administered orally. The cattle showed no signs of clinical toxicity. Among various groups of insects against which insect growth regulators have shown most promise are Diptera of medical and public health significance and investigations on the practical application of methoprene against these targets are still continuing (Chamberlain, 1975).

Control of greenhouse Homoptera Juvenile hormones appear to control larval adult and aptera alate transformation in the aphids (Lees, 1966). Kuhr & Cleere (1973) in their studies on the inhibitory effects of synthetic juvenile hormones on several aphid species report that although the hormone analogues tested did not appear to be highly effective at levels considered to be economically feasible, they had some long term effects such as a wide range of morphological abnormalities, e.g. crumpled wings, inability to give birth on reaching adulthood etc. Exposure to lower levels of the hormone could prove valuable as part of an integrated aphid control programme. However, Staal *et al.* (1973) report that an IGR, kinoprene (2-Propynyl (2E, 4E)-3,7,11-trimethyl-2,4-dodecadienoate), combines a very high and selective JH activity with a direct toxicity on Homoptera at higher doses. Methoprene has also been found to be effective against many species of Homoptera such as aphids, white flies and scales (Staal, 1977).

Insect control in stored commodities Some promising results have been obtained in the use of IGRs against a number of stored product pests (Strong & Dickman, 1973; Hoppe, 1974). Methoprene and hydroprene or Altozar® (ethyl (2E,4E)-3,7, 11-trimethyl-2,4-dodecadienoate), when incorporated into substrates in quantities as small as 5–50 ppm prevented adult emergence of stored product pests such as Coleoptera and Lepidoptera and their performance compared favourably with malathion. However, Staal (1977) reports that the development of IGRs with JH activity has been hampered by the relative insensitivity to these compounds of the two very important stored product pests, i.e. *Sitophilus oryzae* and *S. granarius.*

Control of crop pests Walker (1973) showed that foliage sprays of JHA were effective in the field in preventing the adult development of larvae of the Mexican beetle, *Epilachna varivestis*, and in reducing the fertility of eggs from adult females confined to treated foliage. So far most IGRs with JH activity have been of limited interest in the control of pests of field crops due to their susceptibility to ultraviolet radiation. But Staal (1977) reports the development of a JHA, epofenonane, that combines high activity on several insect species with high stability on foliage.

Control of social insects Vinson & Robeau (1974) reported that IGRs fed via soybean oil to small colonies of the ant, *Solenopsis invicta*, resulted in an immediate drop in pupal numbers due to cannibalism by the workers and a gradual decline in the number of eggs in the colony. Death of colonies occurred at doses of 5 mg or more per colony. ENT-35477, USDA-JH-25, 7-ethoxy-1-(*p*-ethylphenoxy)-3,7-dimethyl-2-octene was found to be the most active IGR in the studies conducted by these authors. Edwards (1975) reports that when laboratory colonies of pharaoh's ant, *Monomorium pharaonis*, were given access only to food containing methoprene, all the brood died in 4–8 weeks and queens ceased to lay eggs about this time. From the above reports IGRs would appear to have real potential for the control of unwanted social insects. Earlier, Cupp & O'Neal (1973) stated that methoprene proved consistently effective, in concentrations of 5 parts per million or more, in disrupting the development, preventing pupation and causing aberrant behaviour in the larvae of imported fire ants, *Solenopsis richteri* and *S. invicta*. Troisi & Riddiford (1974) also found that IGRs, hydroprene and methoprene prevented larval metamorphosis and production of larval broods in *S. invicta*.

Some results of practical significance have also been obtained by the foliar application of an IGR (1-(4-chlorophenyl)-3-(2,6-diflourobenzoyl)-urea) which is a chitin synthesis inhibitor. Turnipseed *et al.* (1974) found it effective against soybean insects. Neal (1974) reported that this compound controlled alfalfa weevil, *Hypera postica*, in small field trials. Taft & Hopkins (1975) reduced the F_1 generation level of the boll weevil, *Anthonomus grandis* by 98% with diflubenzuron mixed with invert sugar or molasses applied as a bait to cotton plants. Other workers (e.g. Ganyard *et al.*, 1977; Lloyd *et al.*, 1977) have obtained even better results in the control of the pest with diflubenzuron. Laboratory studies on the potential of three IGRs including diflubenzuron have shown promise as potential agents for the control of the tsetse fly, *Glossina morsitans morsitans* in the field (Jordan *et al.*, 1979). The discovery of chitin synthesis inhibitors has excited much interest. Such substances interfere with chitin synthesis and kill insects before they mature and reproduce. Since higher animals do not produce chitin, they should be unaffected. Diflubenzuron, a benzoylphenyl urea (Fig. 25) has been extensively studied and is considered to act by inhibiting chitin synthetase, the final enzyme in the pathway by which chitin is

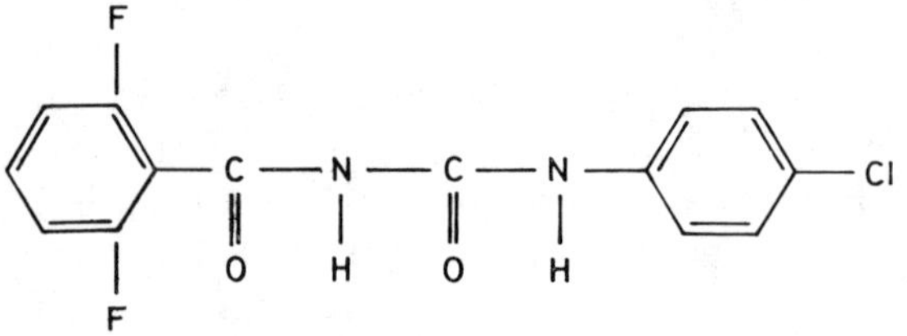

Figure 25 Diflubenzuron.

synthesized from glucose. In the United States, diflubenzuron has already been registered for use against gypsy moth, a serious defoliator of forests. Recently Madrid & Stewart (1981) obtained high mortality in gypsy moth larvae though there was evidence of susceptibility of the parasites in the field on aerial spraying with diflubenzuron. In addition, this IGR has been reported to be effective against immature stages of a broad range of insect species (see Berry *et al.*, 1980 for references). However, it is ineffective against several other insect species, such as cockroaches and ants. Trials on the long term environmental effects of this compound are still continuing.

While diflubenzuron has been found to be very effective against gypsy moth larvae and other caterpillars and its effects on non-target species have been rather minimal (Marx, 1977), the compound does not act immediately. This is because the larvae do not die until they moult. Thus the insecticide must be applied early in the life cycle to minimize damage. For a detailed account of the properties of diflubenzuron, reference should be made to the article by Marx (1977). For a detailed account of the practical results obtained with IGRs, reference should be made to the papers by Staal (1975, 1977).

Anti-hormones Scrutiny of plant extracts in search of anti-hormones has led to the discovery in the common bedding plant, *Ageratum houstonianum*, of two compounds, 7- methoxy and 6,7-dimethoxy-2,2-dimethyl-chromene, named precocene 1 and 2 respectively (Bowers, 1976, 1977) (Fig. 26). Such compounds, also called anti-allatotropins, have been claimed to

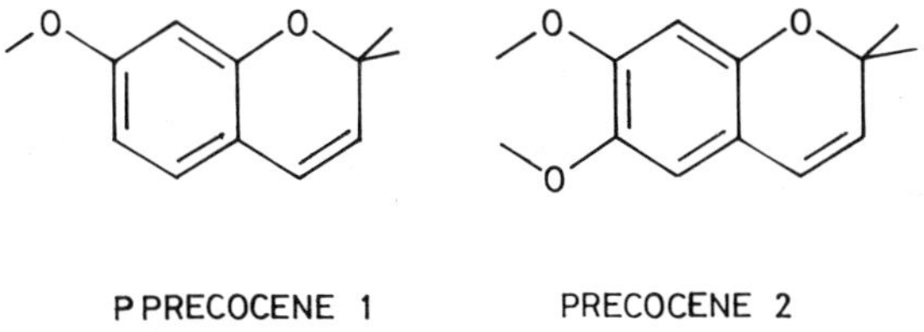

Figure 26 Structure of anti-juvenile hormones extracted from *Ageratum houstonianum.*

represent the fourth generation of insecticides (Bowers *et al.*, 1976; Rondest, 1976). Slåma (1978) has, however, emphasized the great difficulty in recognizing criteria by which to distinguish anti-hormones from various development inhibitions which they induce. According to him, without at least some criteria of specificity, all kinds of growth inhibitors, anti-metabolites, anti-feedants, etc. might be easily confused with one or other type of antihormone.

The so called 'anti-juvenile hormones' or anti-allatotropins belonging to the group of ageratochromenes are mainly effective on the milkweed bug, *Oncopeltus fasciatus*, and cotton stainer bugs, *Dysdercus* spp. (Bowers, 1976). Further, the effects produced by precocene-2 could be counteracted by JH III (Siddall, 1977).

Another group of anti-JH compounds, represented by ethyl-4-2-(tertbutylcarbonyloxy) butoxy benzoate or ETB has been found to affect only the tobacco hornworm, *Manduca sexta.* Staal (1977) also found that piperonyl butoxide (Fig. 26) has all the characteristics of a juvenile hormone antagonist in the *Manduca* test system. In a review of compounds possessing anti-juvenile hormone activity, Staal (1977) has expressed the view that JH antagonists with more useful properties may still await discovery.

Anti-moulting (anti-MH) hormones (e.g. certain azosterols) appear to disrupt the development of phytophagous insects (Robbins *et al.*, 1975). However, caution would be required in the use of such compounds as they appear to have regulatory

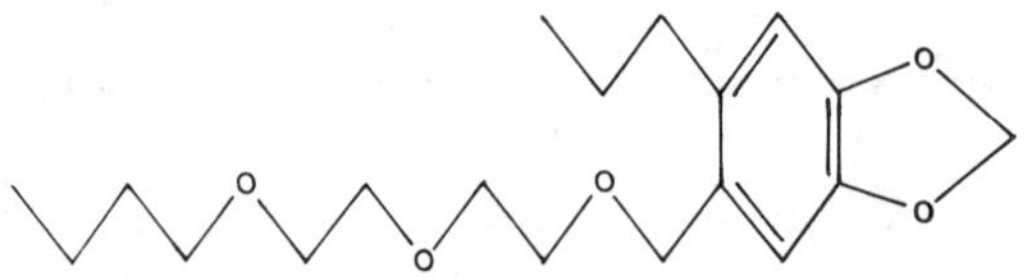

Figure 27 Piperonyl butoxide.

functions in arthropods other than insects as well (Staal, 1977). However, both anti-JH and anti-MH are still largely in the experimental stage.

Future use of insect hormones It had been widely thought that the use of hormones as pesticides would eliminate or reduce the environmental hazards associated with the use of synthetic pesticides. In view of their similarity in physiological action to the naturally-occurring hormones and resemblances in molecular structure, it was considered that juvenile hormones and their analogues would be less liable to induce resistance than conventional insecticides. However, some instances of resistance and a measure of cross tolerance to JH mimics are already known (Cerf & Georghiou, 1972, 1974, Brown & Brown, 1974 and Georghiou, 1975 etc.)

Some disadvantages of insect hormone analogues are as follows (see Menn & Beroza, 1972; Jacobson, 1977; and Staal, 1977 for details):

(1) Lack of specificity.

(2) Slow action and inability to provide immediate pest control and curtail ongoing damage (i.e. plant destruction, disease transmission, etc.) caused by immature and adult stages of insects. These compounds can only be effectively used where preventive control in preceding generations is feasible.

(3) Effective only at certain points in an insect's life cycle which must be monitored to know exactly when to intervene.

(4) Lack of irreversible effects on adults.
(5) Relatively high cost of manufacture.
(6) Short life in the environment.
(7) Lack of activity on some insect groups (grasshoppers, weevils, cockroaches, etc.).
(8) Lack of ovicidal effect in most cases.

Jacobson (1977), in commenting on the future use of growth regulators against insects, thought the outlook uncertain. In addition to the disadvantages mentioned above, only a few areas in plant protection appear amenable to control by these compounds. But in view of their lack of harm to the environment and their effectiveness at low concentrations, vigorous research is being conducted on the possible use of hormones as pesticides. While it appears that attractants, repellents, anti-feedants and insect hormones will not singly solve the problems of economic entomology any more than DDT did, they may be useful in integrated pest management programmes and may stimulate useful research on the control of insect development and differentiation.

12.5 Literature cited

Anonymous (1979). Pheromones of the Sesiidae (formerly Aegeriidae). *Agricultural Research (Northeastern Region), Science and Education Administration, Department of Agriculture, Beltsville, Md. 20705.* 83 pp.

Ascher, K.R.S. & Nissum, S. (1964). Organotin compounds and their potential use in insect control. *World Rev. Pest Control,* **3**, 188–211.

Beevor, P.S. & Campion, D.G. (1979). The field use of 'inhibitory' components of lepidopterous sex pheromones and pheromone mimics, pp. 313–25. In *Chemical Ecology: odour communication in animals.* Ed. F. J. Ritter, Elsevier North-Holland Biomedical Press, Amsterdam.

Bergot, B.J., Jamieson, G.C., Ratcliff, M.A. & Schooley, D.A. (1980). JH Zero: new naturally occurring insect juvenile hormone from developing embryos of the tobacco hornworm. *Science,* **210**, 336–8.

Beroza, M. (ed.) (1970). *Chemicals Controlling Insect Behaviour.* Academic Press, New York. 170 pp.

Beroza, M. (1972). Attractants and repellents for insect pest control, pp. 226–53. In *Pest Control Strategies for the Future*. National Academy of Sciences, Washington, D.C. 376 pp.

Beroza, M. (1976). Control of gypsy moth and other insects with behaviour controlling chemicals, pp. 99–118. In *Pest Management with Insect Sex Attractants.* Ed. M. Beroza. *ACS Symp. Ser.*, **23.** *Am. Chem. Soc. Washington, D.C.* 192 pp.

Berry, E.C., Faragalla, A.A. & Guthrie, W.D. (1980). Field evaluation of diflubenzuron for control of first and second generation European corn borer. *J. Econ. Ent.*, **73(5)**, 634–6.

Borden, J. H. (1977). Behavioral responses of Coleoptera to pheromones, allomones, and kairomones, pp. 169–98. In *Chemical Control of Insect Behaviour.* Eds H. H. Shorey and J. J. McKelvey, Jr. John Wiley and Sons, Inc., New York. 512 pp.

Bottrell, D. R. (1979). *Integrated Pest Management.* Council on Environmental Quality. U.S. Government Printing Office, Washington, D.C. 120 pp.

Bowers, W.S. (1976). Discovery of insect antiallatotropins, pp. 394–408. In *The Juvenile Hormones.* Ed. L. I. Gilbert. Plenum Press, New York. 582 pp.

Bowers, W. S. (1977). Anti-juvenile hormones from plants: chemistry and biological activity. *Pontificiae Academiae Scientiarum Scripta Varia,* **41**, 129–42.

Bowers, W. S., Fales, H. M., Thompson, M. J. & Uebel, E. C. (1966). Juvenile hormone: identification of an active compound from balsam fir. *Science,* **154**, 1020–1.

Bowers, W. S., Ohita, T., Cleere, J. S. & Marsella, P. A. (1976). Discovery of insect antijuvenile hormones in plants. *Science,* **193**, 542–7.

Brown, T. M. & Brown, A. W. A. (1974). Experimental induction of resistance to a juvenile hormone mimic. *J. Econ. Ent.*, **67(6)**, 799–801.

Campion, D. G., McVeig, L. J., Hunter-Jones, P., Hall, D. R., Lester, R., Nesbitt, B.F., Marrs, G. J. & Alder, M. R. (1981). Evaluation of microencapsulated formulation of pheromone components of the Egyptian cotton leafworm, *Spodoptera littoralis* (Boisd.) in Crete. In *Management of Insect Pests with Semiochemicals.* Ed. E. R. Mitchell. Plenum Press, New York. 514 pp.

Campion, D. G. & Nesbitt, B. F. (1981). Lepidopteran sex pheromones and pest management in developing countries. *Tropical Pest Management,* **27(1)**, 53–61.

Carlise, D. B. & Ellis, P. E. (1968). Bracken and locust ecdysones: their effects on moulting in the desert locust. *Science*, **159**, 1472–4.

Cerf, D. C. & Georghiou, G. P. (1972). Evidence of cross-resistance to a juvenile hormone analogue in some insecticide-resistant houseflies. *Nature,* **239**, 401–2.

Cerf, D. C. & Georhgiou, G. P. (1974). Cross resistance to an inhibitor of chitin synthesis, TH 60-40, in insecticide-resistant strains of the housefly. *J. Agr. Food. Chem.*, **22**, 1145–6.

Černy, V., Dolejš, L., Lábler, L., Šorm, F. & Sláma, K. (1967). Dehydrojuvabione, a new compound with juvenile hormone activity from balsam fir. *Collection Czechoslov. Chem. Commun.*, **32**, 3926–31.

Chamberlain, W. F. (1975). Insect growth regulating agents for control of arthropods of medical and veterinary importance. *J. Med. Ent.*, **12**, 395–400.

Chapman, R. F. (1974). The chemical inhibition of feeding by phytophagous insects: A review. *Bull. ent. Res.*, **64**, 339–63.

Chritchley, B. R. (1971). Insect repellents. *PANS*, **17(3)**, 313–14.

Cupp, E. W. & O'Neal, J. (1973). The morphogenetic effects of two juvenile hormone analogues on larvae of imported fire ants. *Environmental Entomology*, **2(2)**, 191–4.

Dethier, V. G., Brown, L. B. & Smith, C. N. (1960). The designation of chemicals in terms of the responses they elicit from insects. *J. Econ. Ent.*, **53**, 134–6.

Drew, R. A. I. (1974). The response of fruit fly species (Diptera: Tephritidae) in the South Pacific area to male attractants. *J. Aust. ent. Soc.*, **13**, 267–70.

Drew, R. A. I., Hooper, G. H. S. & Bateman, M. A. (1978). *Economic Fruit Flies of the South Pacific Region.* Watson Ferguson & Co., Brisbane, Australia. 137 pp.

Edwards, J. P. (1975). The effects of juvenile hormone analogue on laboratory colonies of pharaoh's ant, *Monomorium pharaonis* (L.) (Hymenoptera, Formicidae). *Bull. ent. Res.*, **65**, 75–80.

Ellis, P. (1968). Can insect hormones and their mimics be used to control pests? *PANS* (A), **14**, 329–42.

Fletcher, B. S. (1977). Behavioral responses of Diptera to pheromones, allomones, and kairomones, pp. 129–48. In *Chemical Control of Insect Behaviour.* Eds. H. H. Shorey and J. J. McKelvey, Jr. John Wiley and Sons, Inc., New York. 512 pp.

Flint, H. M., Kuhn, S., Hori, B. & Sallam, H. A. (1974). Early season trapping of pink bollworm with gossyplure. *J. Econ. Ent.*, **67(6)**, 738–40.

Ganyard, M. C., Bradley, J. R., Jr., Boyd, F. J. & Brazzel, J. R. (1977). Field evaluation of diflubenzuron (Dimilin) for control of boll weevil reproduction. *J. Econ. Ent.*, **70**, 347–50.

Georghiou, G. P. (1975). Resistance to insect growth regulators. *WHO, Geneva VBC/EC/75.33.* 16–23 September, 1975, 1–15.

Glass, E. H. (1976). Potential increases in food supply through research in Agriculture: research needs on pesticides and related problems for increased food supplied. (A report to Science and Technology Policy Office, National Science Foundation.) Cornell University, Ithaca, N.Y. 63 pp.

Harris, R. L., Frazer, E. D. & Younger, R. L. (1973). Hornflies, stable flies: development in faeces of bovines treated orally with juvenile hormone analogues. *J. Econ. Ent.*, **66(5)**, 1099–1102.

Hartley, G. S. & Graham-Bryce, I. J. (1980). *Physical Principles of Pesticide Behaviour: the dynamics of applied pesticide in the local environment in relation to biological responses.* Academic Press, London. 2 vols. 1024 pp.

Higgins, R. A. & Pedigo, L. P. (1979). A laboratory antifeedant simulation bioassay for phytophagous insects. *J. Econ. Ent.*, **72(2)**, 238–44.

Hoope, T. (1974). Effect of a juvenile hormone analogue on Mediterranean flour moth in stored grains. *J. Econ. Ent.*, **67(6)**, 789.

Jacobson, M. (1965). *Insect Sex Attractants.* Interscience, New York. 154 pp.

Jacobson, M. (1966). Chemical insect attractants and repellents. *Ann. Rev. Ent.*, **11**, 403–22.

Jacobson, M. (1977). Impact of natural plant protectants on the environment. *Pontificiae Academiae Scientiarvm Scripta Varia,* **41**, 409–30.

Jordan, A. M., Trewern, M. A., Bořkovec, A. B. & DeMilo, A. B. (1979). Laboratory studies on the potential of three insect growth regulators for control of the tsetse fly, *Glossina morsitans morsitans* Westwood (Diptera: Glossinidae). *Bull. ent. Res.*, **69**, 55–64.

Kaae, R. S., Shorey, H. H., McFarland, S. U. and Gaston, L. K. (1973). Sex pheromones of Lepidoptera. XXXVII. Role of sex pheromones and other factors in reproductive isolation among ten species of Noctuidae. *Ann. Ent. Soc. Amer.*, **66**, 444–8.

Kennedy, J. S. (1972). The emergence of behaviour. *J. Aust. ent. Soc.*, **11(3)**, 168–76.

Karlson, P., Hoffmeister, H., Hoppe, W. & Hüber, R. (1963). Zur chemie des ecdysons. *Justus Liebigs Annl.or. Chem.*, **662**, 1–20.

Knox, P. C. & Hays, K. L. (1972). Attraction of *Tabanus* spp. (Diptera: Tabanidae) to traps baited with carbon dioxide and other chemicals. *J. Econ. Ent.*, **1(3)**, 323–6.

Kuhr, R. J. & Cleere, J. S. (1973). Toxic effects of synthetic juvenile hormones on several aphid species. *J. Econ. Ent.*, **66(5)**, 1019–22.

Law, J. H. & Regnier, F. E. (1971). Pheromones. *Ann. Rev. Ent.*, **16**, 533–48.

Lees, A.D. (1966). The control of polymorphism in aphids. *Adv. Insect Physiol.*, **3**, 207–77.

Lloyd, E. P., Wood, R. H. & Mitchell, E. B. (1977). Boll weevil suppression with TH-6040 applied in cottonseed oil as a foliar spray. *J. Econ. Ent.*, **70**, 442–4.

MacDonald, J. F., Akre, R. D. & Hill, W. B. (1973). Attraction of yellowjackets (*Vespula* spp.) to Heptyl butyrate in Washington State (Hymenoptera: Vespidae). *Environmental Entomology,* **2(3)**, 375–9.

Madrid, F. J. & Steward, R. K. (1981). Impact of diflubenzuron spray on gypsy moth parasitoids in the field. *J. Econ. Ent.*, **74(1)**, 1–2.

Madsen, H. F. (1967). Codling moth attractants. *PANS*, **13**, 333–44. pheromone of *Pectinophora gossypiella* (Saund.) (Lepidoptera, Gelechiidae) in Malawi. *Bull. ent. Res.*, **66**, 267–78.

Marks, R. J. (1976). Field evaluation of gossyplure, the synthetic sex.

Marks, R. J. (1977). Assessment of the use of sex pheromone traps to time chemical control of red bollworm *Diparopsis castanea* Hampson (Lepidoptera: Noctuidae) in Malawi. *Bull. ent. Res.*, **67**, 575–87.

Marx, J. L. (1977). Chitin synthesis inhibitors: new class of insecticides. *Science*, **197**, 1170–2.

McLaughlin, J. R., Shorey, H. H., Gaston, L. K., Kaae, R. S. & Stewart, F. D. (1972). Sex pheromones of Lepidoptera. XXXI. Disruption of sex pheromone communication in *Pectinophora gossypiella* with hexalure. *Environmental Entomology*, **1**, 645–50.

Meinwald, J., Prestwich, G. D., Nakanishi, K. & Kubo, I. (1978). Chemical ecology: studies from East Africa. *Science*, **199**, 1167–73.

Meisner, J., Kehat, M., Zur, M. & Eizick, C. (1978). Response of *Earias insulana* Boisd. larvae to Neem (*Azadirachta indica* A. Juss) kernel extract. *Phytoparasitica*, **6**, 85–8.

Menn, J. J. & Beroza, M. (Eds) (1972). *Insect Juvenile Hormones, Chemistry and Action.* Academic Press, New York. 341 pp.

Metcalf, C. L., Flint, W. P. & Metcalf, R. L. (1962). *Destructive and Useful Insects.* 4th Edition. McGraw-Hill, New York. 1087 pp.

Metcalf, R. L. & Metcalf, R. A. (1975). Attractants, repellents, and genetic control in pest management, pp. 275–306. In *Introduction to Insect Pest Management.* Eds R. L. Metcalf, and W. H. Luckman. John Wiley and Sons, New York. 587 pp.

Munakata, K. (1977). Insect feeding deterrents in plants, pp. 93–102. In *Chemical Control of Insect Behaviour.* Eds H. H. Shorey and J. J. McKelvey, Jr. John Wiley and Sons, Inc., New York. 512 pp.

Nakanishi, K. (1977). Insect growth regulators from plants. *Pontificiae Academiae Scientiarum Scripta Varia*, **41**, 185–98.

Neal, J. W., Jr. (1974). Alfalfa weevil control with the unique growth disruptor T.H. 6040 in small plot tests. *J. Econ. Ent.*, **67(2)**, 300–2.

Riddiford, L. M. & Truman, J. W. (1978). Biochemistry of insect hormones and insect growth regulators, pp. 308–357. In *Biochemistry of Insects.* Ed. M. Rockstein. Academic Press, New York. 649 pp.

Robbins, W. E., Thompson, M. J., Svoboda, J. A., Shortino, T. J., Cohen, C. F., Dutky, S. R. & Duncan, O. J. (1975). Nonsteroidal secondary and tertiary amines: inhibitors of insect development and metamorphosis and Δ24-sterol reductase system of tobacco hornworm. *Lipids*, **10**, 353–9.

Rockstein, M. (Ed.) (1978). *Biochemistry of Insects.* Academic Press, New York. 649 pp.

Roelofs, W. L. (Ed.) (1979). Establishing efficacy of sex attractants and disruptants for insect control. *Entomological Society of America, Maryland.* 97 pp.

Roelofs, W. L., Carde, R. T., Taschenberg, E. R. & Weives, R. W., Jr. (1976). Pheromone research for the control of lepidopterous pests in New York. pp. 75–87. In *Pest Management with Insect Sex Attractants.* Ed. M. Beroza. *ACS Symp. Ser., 23. Am. Chem. Soc.,* Washington, D.C. 192 pp.

Roller, H., Dahm, K. J., Sweely, C. C. & Trost, B. M. (1967). The structure of the juvenile hormone. *Angew. Chem. Int. Ed Engl.,* **6**, 179–80.

Rondest, J. (1976). Les antihormones juvéniles: pesticide de demain. *Recherche,* **7**, 975.

Saxena, R. C., Liquido, N. J. & Justo, H. D. (1981a). Neem seed oil, a potential antifeedant for the control of the rice brown planthopper, *Nilaparvata lugens.* pp. 263–77. *In* Natural pesticides from the Neem tree (*Azadirachta indica* A. Juss.). *Proc. 1st Int. Neem Conference, Rottach-Egern,* 1980.

Saxena, R. C., Waldbauer, G. P., Liquido, N. J. & Puma, B. C. (1981b). Effects of Neem seed oil on the rice leaf folder, *Cnaphalocrocis medinalis.* pp. 278–89. *Ibid.*

Schaefer, C. H. & Wilder, W. H. (1972). Insect development inhibitors: a practical evaluation as mosquito control agents. *J. Econ. Ent.,* **65(4)**, 1066–71.

Schooley, D. A. (1977). Analysis of the naturally occurring juvenile hormones – their isolation, identification, and titer determination at physiological levels. pp. 241–287. In *Analytic Biochemistry of Insects.* Ed. R. B. Turner. Elsevier, Amsterdam. 315 pp.

Schooley, D. A., Judy, K. J., Bergot, B. J., Hall, M. S. & Jennings, R. C. (1976). Determination of the physiological levels of juvenile hormones in several insects and biosynthesis of the carbon skeletons of the juvenile hormone, pp. 101–17. In *The Juvenile Hormones.* Ed. L. I. Gilbert. Plenum Press, New York. 582 pp.

Schreck, C. E. (1977). Techniques for the evaluation of insect repellents: a critical review. *Ann. Rev. Ent.* **22**, 101–19.

Shorey, H. H. (1973). Behavioural responses to insect pheromones. *Ann. Rev. Ent.* **18**, 349–80.

Shorey, H. H. (1976). Application of pheromones for manipulating insect pests of agricultural crops, pp. 97–108. In *Insect Pheromones and their Applications.* T. Kono and S. Ishii. Japan Plant Protection Assoc., Tokyo. 179 pp.

Shorey, H. H. (1977*a*). Interaction of insects with their chemical environment, pp. 1–5. In *Chemical Control of Insect Behaviour.* Eds H. H. Shorey and J. J. McKelvey, Jr. John Wiley and Sons, Inc., New York 512 pp.

Shorey, H. H. (1977*b*). Current state of the field use of pheromones in insect control. *Pontificiae Academiae Scientiarum Scripta Varia,* **41**, 385–400.

Shorey, H. H., Gaston, L. K. & Saario, C. A. (1967). Sex pheromones of noctuid moths. XIV. Feasibility of behavioural control by disrupting pheromone communication in cabbage loopers. *J. Econ. Ent.* **69(6)**, 1541–5.

Shorey, H. H., Kaae, R. S. & Gaston, L. K. (1974). Sex pheromones of Lepidoptera. Development of a method for pheromonal control of *Pectinophora gossypiella* in cotton. *J. Econ. Ent.,* **67(3)**, 347–50.

Shorey, H. H. & McKelvey, J. J., Jr. (Eds) (1977). Chemical control of insect behaviour. John Wiley and Sons, Inc., New York. 512 pp.

Siddall, J. B. (1977). Juvenile hormones and their analogs. *Pontificiae Academiae Scientiarum Scripta Varia,* **41**, 37–57.

Sláma, K. (1978). The principles of antihormone action in insects. *Acta entomologica Bohemoslovaca*, **75**, 65–82.

Sláma, K. & Williams, C. M. (1966 *a*). The juvenile hormone. V. The sensitivity of the egg of *Phyrrhocoris apterus* to a hormonally active factor in American paper pulp. *Biol. Bull. mar. biol. Lab. Woods Hole,* **130**, 235–46.

Sláma, K. & Williams, C. M. (1966 *b*). 'Paper factor' as an inhibitor of the embryonic development of the European bug *Phyrrhocoris apterus. Nature*, **210**, 329–30.

Smith, R. L., Flint, H. M. & Forey, D. E. (1978). Air permeation with gossyplure: feasibility studies on chemical confusion for control of the pink bollworm. *J. Econ. Ent.,* **71(2)**, 257–64.

Staal, G. B. (1975). Insect growth regulators with juvenile hormone activity. *Ann. Rev. Ent.,* **20**, 417–60.

Staal, G. B. (1977). Insect control with insect growth regulators based on insect hormones. *Pontificiae Academiae Scientiarum Scripta Varia,* **41**, 353–77.

Staal, G. B., Nassar, S. & Martin, J. W. (1973). Control of the citrus mealybug with insect growth regulators with juvenile hormone activity. *J. Econ. Ent.,* **66**, 851–3.

Steck, W. & Bailey, B. K. (1978). Pheromone traps for moths: evaluation of cone trap designs and design parameters. *Environmental Entomology,* **7(3)**, 449–55.

Strong, R. G. & Dickman, J. (1973). Comparative effectiveness of fifteen insect growth regulators against several pests of stored products. *J. Econ. Ent.,* **66(5)**, 1167–73.

Taft, H. M. & Hopkins, A. R. (1975). Boll weevils: field populations controlled by sterilizing emerging overwintering females with a TH-6040 sprayable bait. *J. Econ. Ent.,* **68**, 551–4.

Tamaki, Y. (1977). Complexity, diversity, and specificity of behaviour modifying chemicals in Lepidoptera and Diptera, pp. 253–85. In *Chemical Control of Insect Behaviour.* Eds H. H. Shorey and J. J. McKelvey, Jr. John Wiley and Sons, Inc., New York. 512 pp.

Teich, I., Neumark, S., Jacobson, M., Klug, J., Shani, A. & Waters, R. M. (1979). Mass trapping of males of Egyptian cotton leaf worm *Spodoptera littoralis* and large scale synthesis of pheromone, pp. 343–50. In *Chemical Ecology: Odour communication in animals.* Ed. F. J. Ritter. Elsevier North-Holland Biomedical Press, Amsterdam. 248 pp.

Trammel, K., Roelofs, W. L. & Glass, E. H. (1974). Sex pheromone trapping of males for control of red banded leafroller in apple orchards. *J. Econ. Ent.,* **67(2)**, 159–64.

Trautmann, K. H., Suchý, M., Masner, P., Wipf, H. K. & Schuler, A. (1976). Isolation and identification of juvenile hormones by means of a radio-active isotope dilution method: evidence for JH III in eight species from four orders. pp. 118–130. In *The Juvenile Hormones.* Ed. L. I. Gilbert. Plenum Press. New York. 582 pp.

Troisi, S. J. & Riddiford, L. M. (1974). Juvenile hormone effects on metamorphosis and reproduction of the fire ant, *Solenopsis invicta. Environmental Entomology,* **3(1)**, 112–16.

Turnipseed, S. G., Heinrichs, E. A., Silva, R. F. P. & Todd, J. W. (1974). Response of soybean insects to foliar applications of a chitin synthesis inhibitor, TH 6040. *J. Econ. Ent.,* **67(6)**, 760–2.

Vinson, S. B. & Robeau, R. (1974). Insect growth regulator effects on colonies of the imported fire ant. *J. Econ. Ent.,* **67(5)**, 584–87.

Walker, W. F. (1973). Mexican bean beetle: compounds with juvenile hormone activity (juvegens) as potential control agents. *J. Econ. Ent.,* **66(1)**, 30–3.

Williams, C. M. (1960). The juvenile hormone. *Acta Endocrinol.* (*suppl.*), **50**, 189–91.

Williams, C. M. (1967). Third generation pesticides. *Sci. Amer.,* **217**, 13–17.

Wright, D. P., Jr. (1963). Antifeeding compounds for insect control. *Adv. Chem. Ser.,* **41**, 56–63.

Wright, D. P., Jr. (1967). Antifeedants, pp. 287–293. In *Pest Control: Biological, physical and selected chemical methods.* Eds W. D. Kilgore and R. L. Doutt. Academic Press, New York. 206 pp.

Wright, J. E., Campbell, J. B. & Hester, P. (1973). Hormones for control of livestock arthropods: evaluation of two juvenile hormone analogues applied to breeding materials in small plot tests in Nebraska and Florida for control of the stable fly. *Environmental Entomology,* **2**, 69–72.

13 Pest Management

13.1 What is and why pest management?

13.1.1 Background information

Pre-DDT era Before the introduction of organochlorine insecticides in the 1940s, the control of pests was mostly labour intensive. It is true that inorganic chemicals, such as Paris green (calcium arsenate) were used and some organic chemicals of plant origin, e.g. nicotine, pyrethrum and rotenone were known as well. But they were seldom applied on as large a scale as some of the modern chemicals and the application technology was in a primitive state. For example, to control locusts and caterpillars, baits were frequently used. However, there was emphasis on cultural practices. Newsom (1974), in his article on the history of pest management in the cotton growing southern United States, has shown that in the pre-calcium arsenate period the farmers were advised to follow some eighteen cultural practices at various intervals. The comprehensive nature of these measures may be judged from Newsom's (1974) statement that 'it seems quite probable that the current pest management programme for cotton insect pests may eventually evolve to the extent that the methods available 45 years ago will, with minor modifications form the core of the pest management systems'. However, with the advent of modern insecticides (See Chapter 10), the farmers discontinued the cultural practices and placed almost exclusive reliance on these toxic substances.

Impact of the development of modern insecticides As noted by Passlow (1976) the results obtained from the use of DDT and other chemicals were spectacular and in the eyes of producers, miraculous. The same period witnessed additional remarkable improvements in machines used for spraying the chemicals. Better pest control was accompanied by improved fertilizer use, weed control by herbicides, irrigation methods and mechanization.

Suddenly, pest control was so easy in terms of application and results. Consequently entomologists were considered redundant (Passlow, 1976). Spraying with insecticides was the answer to pest problems. Biological study was seen as a long-term, largely unnecessary academic exercise. Niceties such as economic relevance, economic threshold, and cultural control were largely ignored by the suddenly successful agricultural community. An insect was a pest and the quickest course was to spray it (Passlow, 1976.).

Following the conventional spraying methods, aerial agriculture was introduced (Passlow, 1976) and as observed by this author the costs were somewhat higher but rapidity of application, lesser dependence on weather factors, lower on-farm labour inputs and convenience relative to other management practices made this technique highly acceptable to the producers. However, the ease of pest control with insecticides was one of the factors responsible for high cost, high return, massive capital input, irrigation production of many crops.

A few entomologists foresaw some of the dangers in virtually total dependence on chemicals but they were ignored (Passlow, 1976). Production and more production per hectare was still the aim of agricultural technology. Very limited finance was available for in depth studies of the basic biology of pests. The very few entomologists involved in field crop studies considered such issues as lower dosages, less comprehensive schedules, the possibilities of resistance, etc. but their voices largely went unheard. The lessons of some well known cases of resistance to pesticides were ignored apparently on the assumption that 'it cannot happen here' (Passlow, 1976). The consequences of the use of insecticides have been fully discussed elsewhere.

13.1.2 Subsequent developments

The history of pest control in the 20th century clearly demonstrates that our present problems of keeping pest populations under control arise from our almost exclusive reliance on one method of control, the use of insecticides. While it is difficult to visualize a time when it will not be necessary to make some use of insecticides for the control of insect pests, the problems entailed in their use, of which the major ones are outlined elsewhere, make it imperative that we alter our present control tactics. As stated by Thomas (1973) 'this is no criticism of pesticides, but a serious question of our intelligence in using them . . . we must be aware of the concerns of environmentalists and ecologists and their demand that we justify each and every use of pesticides'.

The magnitude of our pest problem is vast and a solution to them undoubtedly lies in our ability to understand each problem and devise a solution acceptable in terms of our overall economic, environmental and social gains. Already examples are at hand where implementation of large scale pest management programmes, for example, in the U.S.A., has cut down the use of insecticides by more than 50% on such crops as cotton, tobacco, sorghum, peanuts, oranges and grapes similar approach has resulted in a drastic reduction in the use of insecticides on cotton in Central America (Table 11). These programmes are based on a sound basic knowledge of the ecology and behaviour of the pests and have had to be developed locally for each pest. This approach has been termed *Integrated Pest Management* (*IPM*).

13.2 Definition of pest management

The fourth session of the FAO Panel of experts (Anonymous, 1972) defined *Pest Management* as, 'an all inclusive term that describes man's continuous efforts to control populations of pest species at levels that are advantageous to his well being'. They also added 'although there has been a widespread tendency to synonymize the terms pest management and integrated pest control, they are not synonymous. Pest management includes all approaches ranging from a single control method, i.e. the repetitive application of a broad spectrum insecticide schedule without regard to population densities or economic injury levels, to the most sophisticated integrated control systems. Thus pest management is a general term which applies to any form of pest population manipulation invoked by man, its objective being to optimize

Table 11 A summary of insecticide usage for the control of cotton pests in Central America (after L.A. Falcon, personal communication).

	Pest Control Programmes		
		Integrated Control Approach	
Pest Problem	**Chemical Approach**	**Early**	**Advanced**
Boll weevil	15–20 applications, 4 days intervals when plants 21 days old	5–10 applications, based on need as determined by scouting	Use trap crop, handpick or, apply chemical pesticide only to trap crop
Cotton leafworm	5–10 applications during first 100 days of cotton plant growth	0–3 applications of selective pathogens (Bt) as determined by scouting	Same as for Early
Bollworm	10–20 applications, especially during boll formation period	5–10 applications as determined by scouting	0–5 applications of selective virus as determined by scouting
Other pests	5–10 applications, when present	0–5 as determined by scouting	0
Total applications	25–50	8–25	0–5
Kg of pesticide (a.i) per metric ton of cotton produced	50–100 kg	20–50 kg	0–1
Cotton production ha^{-1}	2–5 bales actual	2.5–5.5 bales actual	3–6 bales estimated

control in terms of overall economic, social and environmental needs of mankind'.

Integrated control has been comprehensively defined by Smith & van den Bosch (1967) as 'a pest population management system that utilized all suitable techniques either to reduce pest populations and maintain them at levels below those causing economic injury or to so manipulate the populations that they are prevented from causing such injury. Integrated control achieves this ideal by hormonizing techniques in an organized way by making the techniques compatible, and by blending them into a multifaceted flexible system'.

The United States Council on Environmental Quality in their publication *Integrated Pest Management* (November, 1972) defines integrated pest control as 'an approach that employs a combination of techniques to control the wide variety of potential pest that may threaten crops'. They added 'it involves maximum reliance on natural pest population controls, along with a combination of techniques that may contribute to suppression . . . cultural methods, pest-specific diseases, resistant crop varieties, sterile insects, attractants, augmentation of parasites or predators, or chemical pesticides as needed'.

Hogan (1973) believes that in its complete form, integrated control involves 'the collection of appropriate data to enable the construction of a model from which, amongst other things, the effect of population suppression treatments can be predicted and their most efficient usage defined'. Thus, if we are using insecticides, then it should be possible to predict their effect on a particular agro-ecosystem.

The Entomological Society of America Publication 75–2 (Glass, 1975) uses the hybrid term *Integrated Pest Management* and defines it as a, 'pest management system that, in the context of the associated environment and the population dynamics of the pest species, utilizes all suitable techniques and methods in as compatible a manner as possible and maintains pest populations at levels below those causing economic injury'.

Considered critically, both integrated control and pest management may be equated with what Hansberry (1968) called 'good entomology'. Both have the same end results in view, i.e. the development of pest control methods consistent with our realistic ecological and economic objectives. Indeed the two terms are now considered more or less synonymous. Thus Solomon (1973) acknowledges with some reservations the fact that the concept of integrated control seems to be increasingly identified with pest management. At the December 1972 F.A.O. conference in Rome, it was agreed, partly due to the difficulty of translating pest management into some languages, that the term 'integrated control' should be the accepted term. Both pest management and integrated control are widely used in English communications and should be considered as synonymous. Further, Way (1974) made the realistic statement that pest management is a 'generalized, all-embracing term virtually synonymous with pest control'.

The subject of pest management is clearly in its 'infancy' and 'its concepts and strategies have only recently begun to evolve' (Thomas, 1972). This is especially the case in nearly all African countries. But the term pest management emphasizes a shift in our pest control philosophy. Our basic aim remains the minimizing and prevention of losses caused by pests. However, the tools used for this purpose are now more in number and some quite complex in their operations. Pest management systems offer economic and environmental benefits to the producer and to society but require for their development, a sound basic knowledge of the pest and its environment. In practice, the concept of pest management is reduced to the identification of the key factors responsible for the development of insect populations to pest proportions and their manipulation in such a way as to minimize pest damage consistent with our earlier mentioned goals (see Rabb *et al.*, 1974 for a discussion on manipulating the agro-ecosystem). As stated earlier such programmes would have to be developed locally for each pest. For this reason, for the developing world, the days of effective control of pests by purchasing insecticides and application technology in the international market are fast running out. One has only to consider some of the steps involved in developing meaningful pest management programmes (see Smith, 1968 and Bottrell, 1979 for guidelines for the development of IPM; Jones, 1968 for sociology and Pimental, 1978 for legal aspects);

(1) Identification of the pest and a study of its general biology and behaviour; original wild and alternative host(s) to be determined.

(2) Construction of life tables of the crop with a view to obtaining data on the losses as a result of infestation. Determination of accurate economic threshold infestation levels for each pest with a view to deciding when to control the pest.
(3) Construction of life tables of the insect pest and determination of the mortality factors by a simple regression analysis. This would provide an idea of the role of natural enemies in the regulation of the pest population as well as a rough evaluation of the pest population as well as the times and places of their occurrence.
(4) To determine if the pest build up is in some ways influenced by environmental factors. If yes, then manipulate these factors in a plan of pest management with a view to preventing the pest from causing economic damage.
(5) To obtain information on the impact of pesticides on natural enemies with a view to avoiding spraying during the periods of their abundance.
(6) Determination of cost/potential benefit ratio of the intended control measures with a view to deciding on the selection of best remedial measures.
(7) Development of reliable pest prediction methodologies with a view to applying control measures when they are likely to be most effective. Emphasis must be on controlled use of insecticides.
(8) Development of high level of scientific background on which long-term sound pest management programmes can be based. It must be borne in mind that such programmes require ecological analysis, interpretation, and continued monitoring and identification of the key factors which should be manipulated to minimize pest damage.

Smith (1972) emphasizes the existence of a special opportunity to develop pest management systems in the less developed countries of the tropics and subtropics. He rightly argues that in contrast to the temperate regions, physical environment in these areas is, in general, conducive to population increase of natural enemies. Further, most ecosystems in the developing world have, comparatively been less subjected to intensive application of insecticides. Way (1973) has stressed the necessity for basic and 'explanatory' research for developing satisfactory practices for the management of pests. Novel ideas, fresh approaches to complex problems as required for the solution of pest problems can only be arrived at by planned and continuing research, and nowhere is the need greater than in the African countries (Kumar, 1980).

13.3 National/international pest management

Losses due to pests are ultimately paid for by the community in higher food prices. The cost of research and development of new chemicals is financed by the commercial organization concerned but recouped from the consumer. Because of its responsibility to society, the government is at various levels concerned with pest control, for food and health are primary requirements of society. Commercial organizations have, in the past and continue even now, to play a significant role in the control of pests.

The cost of such control measures as quarantine, biological control, and research on pests is beyond the means of individual or even groups of well-to-do farmers and must usually be met by the government. Neighbouring governments may collaborate on a regional basis to control pests for the welfare of their people. It is only an organized authority such as a government agency which can, through its quarantine measures, prevent a pest from being introduced to a country or, if it has entered, to eradicate it before it is firmly established. It is only a government organization or a similar institution which can sponsor basic research which may ultimately develop novel pest management schemes. At this stage it seems worth considering in some detail certain components of pest management systems which need to be dealt with at national/international level.

13.3.1 Some components of pest management systems at national/international level

Quarantine regulations This is the first line of defence against pests and is obviously an important component of pest management systems. It requires co-ordination at national and international levels. Available statistics show that plant and animal pests and diseases cost the United States $12 billion in

1976 and many of their most destructive pests are foreigners that cause greater damage in the U.S.A. than in their natural environment where natural enemies hold them in check. Not long ago, the Mediterranean fruit-flies (*Ceratitis capitata*) were discovered in the U.S.A. on four different occasions and had to be eradicated at a cumulative cost of over $20 million. But the decision to control them saved the citrus industry. In 1971, an imported, diseased parakeet was responsible for an outbreak of exotic Newcastle disease in U.S. poultry. More than 11 million exposed chickens were destroyed to control the deadly bird virus. Commercial poultrymen received over $26 million in compensation indemnities.

In Australia, plant quarantine under Commonwealth legislation has a major responsibility in keeping out of Australia unwanted pests. It is one of the most alert services in the world, with the responsibility to foresee where pest risks may arise, and must be ready to adopt new techniques and procedures in the light of developments both in science and trade. The cost of the plant quarantine programme is of the order of $A600 000 a year (1967 figures). This is a small price to pay for protection when one considers that Australia's export of primary products in 1963–64 earned more than $A1800 million and the national farm income for the same period was nearly $A1200 million. Great Britain spends about £50 000 a year to enforce quarantine laws against the Colorado beetle (*Leptinotarsa decemlineata*). This is a good investment compared with the losses that would follow establishment of the beetle on the island.

Effective, plant quarantine rests on the following fundamental pre-requisites:

(1) The most important requirement is to identify the insect accurately and quickly and recognize that it is an alien to the country. The quarantine measures must be based on sound biological grounds and a thorough knowledge of the pest which is intended to be kept out of the country.

(2) Before a quarantine prohibition or restriction is recommended to the government, the subject requires careful and thorough investigation and reliable advice should be sought on the insects which are to be kept out of the country.

(3) Quarantine must derive legal authority and must operate within the provisions of such law.
(4) The service must be manned by professionally competent officers who keep abreast of pest problems. The importance of correct identification of plants and animals cannot be over-emphasized. In the U.S.A., Khapra beetle (*Trogoderma granarium*) from India gained entry and established itself in a California warehouse. There it was mistaken for the domestic carpet beetle. The United States department of Agriculture has expended over $11 million in eradication measures against this one beetle alone.

In Africa, the Inter-African Phytosanitary Organization, with headquarters in Cameroun, has recently become active. Its thirteenth meeting (Anonymous, 1977) made a series of important recommendations on checking the spread of new pests and diseases in Africa. It will, however, have little impact on the pest situation in the continent unless it has imaginative programmes based on scientific research and the national governments give it their full support and co-operation.

Biological control Previously it was emphasized that good pest management systems endeavour to maximize the effect of natural enemies on pest populations. Biological control in the classical sense means the utilization of parasites, predators and pathogens for the control of pests. However, in recent years biological control has assumed a much broader meaning. In addition to the natural enemies, the following pest suppression measures, termed 'parabiological methods' by Sailer (1976) are included under biological control:

Autocidal methods, principally the sterile male technique, the use of sex attractants, chromosomal translocations, cytoplasmic incompatibility; use of growth regulators and plant resistance.

All these methods involve considerable expenditure of research effort and often require co-ordinated multidisciplinary approaches. The initial cost of implementing biological control measures involving natural enemies is obviously rather high as natural enemies have to be found and collected, often in foreign countries (see Chapter 8).

Operational aspects of exploration for parasites in foreign lands are discussed by Sailer (1974). To give an idea of the complexity of the operation, characteristics and qualifications of personnel required for this type of work, according to Sailer (1974), are as follows:

(1) Excellent training in entomology.
(2) Sound background in ecology.
(3) Broad knowledge of American agricultural practices and pest problems.
(4) Proficiency in 1 and preferably 2 major European language (fluency in English assumed).
(5) No family responsibilities.
(6) Physical stamina.
(7) Ability to deal effectively with foreign farmers, scientists and bureaucrats of the homegrown and foreign varieties.
(8) High frustration tolerance.

In order to develop ecologically sound pest management systems, we require many more professional workers and a greatly expanded programme of exploration of foreign enemies, augmentation of local enemies and attempts to introduce germ plasm. Although investment in these uncertain ventures appears high, returns so far in terms of protecting crops (Chapter 8) show that they compete very favourably with chemical methods and have the added advantage of safeguarding the environment.

Pest assessment and forecasting Programmes in this category are largely financed by governments but organizations of producers may undertake some of these tasks and provide useful services.

The dynamics of pest problems were discussed previously and the need to monitor regularly the pest intensity/crop loss was emphasized. Assessment of the importance of a pest can be accurately done only from widespread observations continuing over several years. Once again, most food exporting countries assess pests by routine estimates of annual densities at many different sites. Frequently there are attempts to determine the relationship between pest intensity and environmental parameters with a view to pest forecasting. This is a field of research where investments and co-operation at national levels are required to develop devices, from simple empirical equations to complex mathematical models, and considerable progress in this direction has been made in Japan, Europe and the U.S.A. In Africa, services to forecast the outbreaks of armyworm and locust attacks exist but little to monitor other pests.

13.3.2 Advantages of adoption of national/international pest management strategies (modified from Southwood & Norton, 1973)

As noted by Southwood & Norton (1973), at least theoretically, there is much to be gained by adopting pest control strategies at regional as opposed to individual farm levels. At least the following advantages may be recognized:

(1) The use of control measures requiring wide geographic applications, for example biological control schemes must be conducted over large areas and require facilities and resources not available to a farmer.

(2) Management systems over large areas usually provide highly favourable cost/benefit ratios.

(3) National strategies may persuade the farmers to adopt certain control methods which are in the interest of the country's long term environmental and efficiency goals.

In the advanced countries of the West which are generally food exporters, the national and state governments are closely involved with research and extension of pest management systems. In Australia, for example, the Commonwealth Scientific and Industrial Research Organization (C.S.I.R.O.), Department of Primary Industries (D.P.I.) and the universities are all concerned with work on pests. C.S.I.R.O. is largely involved in mission oriented basic research on pests while the universities have deeper commitments to fundamental research and the important role of training personnel for C.S.I.R.O. and D.P.I. In addition to applied research, D.P.I. must provide solutions to day-to-day pest problems as well as carrying out the necessary extension duties. In the United Kingdom, the National Agricultural Advisory Service has units throughout the country. Besides there are organizations such as the Tobacco and the Pea Growers' Association which give advance warnings of pest attacks to the growers. In the United States of America, the Department of Agriculture justifiably boasts that as a result of the research and extension efforts of its staff, Americans spend only 15% of their income on food. The department strives to maintain a balanced programme on pests and their management problems in its research, development, education, regulatory and action plans (Cutler, 1978).

13.4 Role of extension services

To what extent the pest management strategies can be implemented depends how effectively the ideas can be conveyed to the farmers in the field. In this context local sociological conditions must be borne in mind. For example, it should be remembered that agriculture in Africa has certain distinctive characteristics. Here, shifting cultivation is widely practised, mixed cropping is common, farms are small and scattered, farming is labour intensive and capital scarce, and the educational background of the farmers is generally very poor.

In most parts of the tropics, the small farmer is at present and is likely in the future to be responsible for the greater part of agricultural production. According to Brader (1979) integrated pest management approach offers the best solution to pest problems under the condition mentioned above. Though it must be realized that large monocultures of wheat and rice, without significant pest management effort, are successfully working in many countries of South and South-East Asia and, nearer home, Sudan has converted the barren desert and grass savanna into a vast cotton growing area. But these changes have been brought about by widespread changes in agricultural practices including multiple cropping, irrigation and increasing use of fertilizers and pesticides. In most countries of the third world however, pest management strategies must in their broadest sense relate to the needs of the small farmer (see Farrington, 1977 for some aspects of research-based recommendations versus farmers practices). For example, some 20 million small farmers live in West Africa, 14 million in Nigeria alone. The need for efficient extension services to reach these farmers cannot be over-emphasized. The value of control measures may be lost unless applied at the right rate and the proper time. Extension services are often the most neglected aspect of crop protection programmes of developing countries in spite of being the vital element providing the feedback through the various links to policy makers. To ensure that the farmer receives maximum benefit from research, he must have easy access to the extension service of the ministry of agriculture or to specialist organizations. This requires a major effort by governments to train a large body of suitable manpower to reach the small farmer. Further pest management must itself be seen in the wider perspective of agricultural improvement including farming system, new varieties, fertilizers and farmers' education. Perrin (1977) has attempted to highlight the urgent need for further involvement of entomologists in the multi-disciplinary approach to multiple cropping research in the tropics. A similar view has been expressed by Litsinger & Moody (1976). With diversity of crops and pests mentioned earlier (see Chapter 2), the immense potential for manipulating the environment in diverse parts of Africa to keep pests in check remains unrealized and must be regarded as an example of African under-development. To this must also be added the fact that many available research results are often not applied through failure of extension effort.

Pest management implementation is essentially an educational process and should be the responsibility of departments of agriculture and health at all levels – villages, small towns and cities. This calls for a complete reorganization of the present pest control services, if they already exist, with careful planning, adequate resources, educational and administrative support. There is a clear need of establishing a cadre of Pest Management Extension Specialists, right from the top specialist down to trainee specialist. There is the need for development of well illustrated public educational materials on the principles of integrated pest management for these personnel. Books and other authoritative materials for the educators are also required. Several books relating to insect pest management in north America and Europe are now available (see, for example, Metcalf & Luckman, 1975; Smith & Pimental, 1978, etc.) but similar manuals on African insect pests are as yet to appear. The Council on Environmental Quality of the United States has produced a well reasoned text (Bottrell, 1979) on Integrated Pest Management outlining its principles, guidelines, control techniques, major achievements in the United States as well as barriers to its progress.

Publications of this kind are sorely required in African countries where paucity of such works may partly be due to the fact that there has so far been little basic work on pest management strategies, for local pests.

13.5 Literature cited

Anonymous (1972). Report of the Expert Panel on Integrated Control. Dec. 1972. *F.A.O., Rome, Italy.*

Anonymous (1977). Report and recommendations. Fifth meeting held under the charter of the Organization of African Unity. *13th Meeting of the Inter-African Phytosanitary Council, Accra.* Mimeographed.

Bottrell, D. R. (1979). *Integrated Pest Management.* Council on Environmental Quality. U.S. Government Printing Office, Washington, D.C. 120 pp.

Brader, L. (1979). Integrated pest control in the developing world. *Ann. Rev. Ent.,* **24**, 225–54.

Brown, A.W.A. (1960). Mechanisms of resistance against insecticides. *Ann. Rev. Ent.,* **5**, 301–26.

Cutler, M.R. (1978). The role of USDA in integrated pest management, pp. 9–20. In *Pest Control Strategies.* Eds E. H. Smith, and D. Pimental. Academic Press, New York. 334 pp.

Farrington, J. (1977). Research based recommendations versus farmers' practices: some lessons from cotton-spraying in Malawi. *Expl. Agric.,* **13**, 9–15.

Glass, E. H. (1975). Integrated pest management: Rationale, potential, needs and implementation. *E. S. A. Special Publication,* 75–2, 141.

Hansberry, R. (1968). Prospects for nonchemical insect control – an industrial view. *Bull. Ent. Soc. Amer.,* **14(3)**, 229–35.

Hogan, T. W. (1973). The integration process as it affects Entomology. *J. Aust. ent. Soc.,* **12**, 241–47.

Jones, D. P. (1968). Integrated control of pests. *PANS,* **14(4)**, 514–22.

Kumar, R. (1980). Research in tropical insect science: a Ghanaian perspective. *An Inaugural Lecture delivered on 7th February, 1980. Ghana Universities Press, P.O. Box 4219, Accra.* 18 pp.

Litsinger, J.A. & Moody, K. (1976). Integrated pest management in multiple cropping systems. Chapter 15. pp. 293–316. In *Multiple Cropping.* Eds R. I. Papendick, P. A. Sanchez, and G. B. Triplett. *Am. Soc. Agron. Madison, Wisconsin.* Special Public. No 27.

Metcalf, R. L. & Luckman, W. H. (Eds) (1975). *Introduction to Insect Pest Management.* John Wiley and Sons, New York. 587 pp.

Newsom, L. D. (1974). Pest management: history, current status and future progress, pp. 1–18. In *Proceedings of the Summer Institute on Biological Control of Plant Insects and Diseases.* Eds F. G. Maxwell, and F. A. Harris. University Press of Mississippi, Jackson. 647 pp.

Passlow, T. (1976). Some aspects of field crop entomology. *News Bull. Ent. Soc. Qld.,* **4(1)**, 6–13.

Perrin, R. M. (1977). Pest management in multiple cropping systems. *Agro-Ecosystems,* **3**, 93–118.

Pimental, D. (1978). Socio-economic and legal aspects of pest control, pp. 55–71. In *Pest Control Strategies.* Eds E. H. Smith and D. Pimental. Academic Press, New York. 334 pp.

Rabb, R. L., Stinner, R. E. & Carlson, G. A. (1974). Ecological principles as a basis for pest management in the agroecosystem, pp. 19–45. In *Proceedings of the Summer Institute of Biological Control of Plant Insect and Diseases.* Eds F. G. Maxwell and F. A. Harris. University Press of Mississippi, Jackson. 647 pp.

Sailer, R. I. (1974). Foreign Exploration and importation of exotic and arthropod parasites and predators, pp. 97–108. In *Proceedings of the Summer Institute on Biological Control of Plant Insects and Diseases.* Eds F. G. Maxwell and F. A. Harris. University Press of Mississippi, Jackson. 647 pp.

Sailer, R. I. (1976). Future role of biological control in management of pests. *Proc. Tall. Timbers Conf. Ecol. Anim. Control Habitat Management,* 195–209.

Smith, R. F. (1968). Recent developments in integrated control. *PANS.* **14(2)**, 201–6.

Smith, R. F. (1972). The impact of the green revolution on plant protection in tropical and subtropical areas. *Bull. Ent. Soc. Amer.* **18(1)**, 7–14.

Smith, E. H. & Pimental, D. (1978). *Pest Control Strategies.* Academic Press, New York. 334 pp.

Smith, R. F. & van den Bosch (1967). Integrated control, pp. 295–340. In *Pest Control – biological, physical and selected chemical methods.* Eds W. W. Kilgore and R. L. Doutt. Academic Press, New York. 206 pp.

Solomon, M. E. (1973). Ecology in relation to the management of insects. In *Insects: studies in population management.* pp. 153–67. Eds P. W. Geir, L. R. Clark, D. J. Anderson and H. A. Nix, *Mem. Ecol. Soc. Australia, Canberra,* **1**, 295 pp.

Southwood, T. R. E. & Norton, G. A. (1973). Economic aspects of pest management strategies and decisions. pp. 168–95. In *Insects: studies in population management.* Eds P. W. Geir, L. R. Clark, D. J. Anderson and H. A. Nix, *Mem. Ecol. Soc. Australia, Canberra,* **1**, 295 pp.

Thomas, J. G. (1972). Pest management implementation. In *Integrated Pest Management, United States Council on Environmental Quality, Washington, D.C.*

Thomas, J. G. (1973). *Implementation of Practical Pest Management Programs.* North Central Branch Meeting of ESA, March 29, 1973, 1–15.

Way, M. J. (1973). Objectives, methods and scope of integrated control. In *Insects: studies in population management.* pp. 135–52. Eds P. W. Geir, L. R. Clark, D. J. Anderson and H. A. Nix, *Mem. Ecol. Soc. Australia, Canberra,* **1**, 295 pp.

Way, M. J. (1974). Integrated control in Britain, pp. 196–208. In *Biology in Pest and Disease Control.* The 13th symposium of the British Ecological Society, Oxford, 4–7 January, 1972. Eds J. D. Price and M. E. Solomon. Blackwell Scientific Publications, Oxford. 408 pp.

Author Index

Bibliographical citations are in *italics*.

Subject Index

The titles of papers cited in the bibliographies and table contents are not indexed.